Patterns

Functions, Reasoning, and Problem Solving

Teacher's Guide

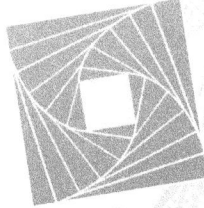

This material is based upon work supported by the National Science Foundation under award numbers ESI-9255262, ESI-0137805, and ESI-0627821. Any opinions, findings, and conclusions or recommendations expressed in this publication are those of the authors and do not necessarily reflect the views of the National Science Foundation.

Key Curriculum
1150 65th Street
Emeryville, California 94608
email: editorial@keypress.com
www.keycurriculum.com

First Edition Authors
Dan Fendel, Diane Resek, Lynne Alper, and Sherry Fraser

Contributors to the Second Edition
Sherry Fraser, Jean Klanica, Brian Lawler, Eric Robinson, Lew Romagnano, Rick Marks, Dan Brutlag, Alan Olds, Mike Bryant, Jeri P. Philbrick, Lori Green, Matt Bremer, Margaret DeArmond

Project Editors
Joan Lewis, Sharon Taylor

Consulting Editor
Mali Apple

Editorial Assistant
Juliana Tringali

Professional Reviewer
Rick Marks, Sonoma State University

Calculator Materials Editor
Christian Aviles-Scott

Math Checker
Christian Kearney

Production Director
Christine Osborne

Executive Editor
Josephine Noah

Textbook Product Manager
Timothy Pope

Publisher
Steven Rasmussen

Contents

Blackline Masters

Lonesome Llama Blackline Master
1-Inch Graph Paper Blackline Master
1/4-Inch Graph Paper Blackline Master
1-Centimeter Graph Paper Blackline Master
In-Class Assessment
Take-Home Assessment

Calculator Guide and Calculator Notes

Introduction

Patterns Unit Overview

Intent

This unit is an essential introduction for students to the variety of ways for working on and thinking about mathematical problems. Students are introduced to general learning skills and methods that are developed and used throughout the four-year IMP curriculum and that form the foundation of the learning process through which students will build mathematical ideas. Here is a summary of these learning skills and methods.

- Working in groups to analyze problems
- Learning about group cooperation and roles in group learning
- Expressing mathematical ideas orally and in writing
- Making presentations in small groups and to the class
- Developing strategies for solving problems
- Using concrete mathematical models
- Doing investigations in which the task is not clearly defined
- Becoming familiar with alternative forms of assessment, such as self-assessment and portfolios
- Learning about the use of graphing calculators and, if available, appropriate computer software

Mathematics

Patterns emphasizes extended, open-ended exploration and the search for patterns. Important mathematics introduced or reviewed in *Patterns* include In-Out tables, functions, variables, positive and negative numbers, and basic geometry concepts related to polygons. Proof, another major theme, is developed as part of the larger theme of reasoning and explaining. Students' ability to create and understand proofs will develop over their four years in IMP; their work in this unit is an important start. This unit focuses on several mathematical ideas:

- Finding, analyzing, and generalizing geometric and numeric patterns
- Analyzing and creating In-Out tables
- Using variables in a variety of ways, including to express generalizations
- Developing and using general principles for working with variables, including the distributive property
- Working with order-of-operations rules for arithmetic

- Using a concrete model to understand and do arithmetic with positive and negative integers

- Applying algebraic ideas, including In-Out tables, in geometric settings

- Developing proofs concerning consecutive sums and other topics

Read "Patterns: A Mathematical Commentary" by Eric Robinson, Professor of Mathematics at Ithaca College, Ithaca, New York.

Progression

In *The Importance of Patterns,* the unit opens with an introduction to functions and their representations. Students begin to build their ability to tackle novel mathematical problems and have their first experiences with graphing calculators. The activities in *Communicating About Mathematics* build on the strands begun in *The Importance of Patterns,* focusing on the written and oral communication of mathematical ideas. In *Investigations,* students explore several rich mathematical tasks while employing the tools and techniques they have developed so far. Finally, in *Putting It Together,* students bring all of their new mathematical tools and techniques, as well as their developing identity as a learning community, to bear on a group of summary activities.

The Importance of Patterns

Communicating About Mathematics

Investigations

Putting It Together

Supplemental Activities

Unit Assessments

Pacing Guides

50-Minute Pacing Guide (22 days)

Day	Activity	In-Class Time Estimate
The Importance of Patterns		
1	Welcome students	10
	What's Next?	30
	Homework: *Past Experiences*	10
2	Discussion: *Past Experiences*	10
	POW 1: The Broken Eggs	35
	Homework: *Who's Who?*	5
3	Discussion: *Who's Who?*	15
	The Standard POW Write-up	20
	Homework: *Inside Out*	15
4	Discussion: *Inside Out*	45
	Homework: *POW 1: The Broken Eggs*	5
5	Discussion: *POW 1: The Broken Eggs*	10
	Calculator Exploration	40
	Homework: *Pulling Out Rules*	0
6	Discussion: *Pulling Out Rules*	20
	Lonesome Llama	30
Communicating About Mathematics		
	Homework: *POW 1: The Broken Eggs* (continued)	0
7	Discussion: *POW 1: The Broken Eggs* (continued)	5
	Role Reflections	15
	Marcella's Bagels	30
	Homework: *POW 1: The Broken Eggs* (continued)	0
8	Discussion: *POW 1: The Broken Eggs* (continued)	15
	1-2-3-4 Puzzle	30
	Homework: *Uncertain Answers*	5
9	Discussion: *Uncertain Answers*	15

	Extended Bagels	15
	Discussion: Expectations for presentations	15
	Homework: *POW 1: The Broken Eggs* (continued)	5
10	Presentations: *POW 1: The Broken Eggs*	15
	The Chefs' Hot and Cold Cubes	35
	Homework: *Do It the Chefs' Way*	0
11	*POW 2: Checkerboard Squares*	10
	Discussion: *Do It the Chefs' Way*	30
	Reflection: First 11 days of class	5
	Homework: *You're the Chef*	5

Investigations

12	Discussion: *You're the Chef*	10
	Consecutive Sums	30
	Homework: *Add It Up*	10
13	Discussion: *Add It Up*	10
	Consecutive Sums (continued)	30
	Homework: *Group Reflection*	10
14	*Consecutive Sums* (continued)	25
	Discussion: *Group Reflection*	15
	Discussion: *POW 2: Checkerboard Squares* (continued)	10
	Homework: *That's Odd!*	0
15	Discussion: *That's Odd!*	10
	Pattern Block Investigations	40
	Homework: *Degree Discovery*	0
16	Discussion: *Degree Discovery*	15
	Polygon Angles	30
	Homework: *An Angle Summary*	5
17	Discussion: *An Angle Summary*	10
	Presentations: *POW 2: Checkerboard Squares*	20

Putting It Together

	Squares and Scoops	20
	Homework: *Another In-Outer*	0
18	Discussion: *Another In-Outer*	30
	Homework: *Diagonally Speaking*	20
19	Discussion: *Diagonally Speaking*	20
	The Garden Border	25
	Homework: *Border Varieties*	5
20	Discussion: *Border Varieties,* including equivalent expressions and the distributive law	30
	Homework: *Patterns* Portfolio	20
21	*In-Class Assessment*	40
	Homework: *Take-Home Assessment*	10
22	Discussion: *In-Class Assessment* and *Take-Home Assessment*	30
	Unit Reflection	20

90-minute Pacing Guide (13 days)

Day	Activity	In-Class Time Estimate
The Importance of Patterns		
1	Welcome students	10
	What's Next?	30
	POW 1: The Broken Eggs	30
	The Standard POW Write-up	15
	Homework: *Who's Who?* and *Past Experiences*	5
2	Discussion: *Who's Who?* and *Past Experiences*	30
	Inside Out	60
	Homework: *POW 1: The Broken Eggs*	0
3	Discussion: *POW 1: The Broken Eggs*	10
	Calculator Exploration	40
	Lonesome Llama	30
	Role Reflections	5
	Homework: *Pulling Out Rules*	0
	Homework: *POW 1: The Broken Eggs* (continued)	5
4	*Role Reflections* (continued)	15
	Discussion: *Pulling Out Rules*	20
Communicating About Mathematics		
	Marcella's Bagels	30
	Discussion: *POW 1: The Broken Eggs* (continued)	20
	Homework: *POW 1: The Broken Eggs* (continued)	5
5	Discussion: Expectations for presentations	15
	Presentations: *POW 1: The Broken Eggs*	20
	Extended Bagels	15
	The Chefs' Hot and Cold Cubes	35
	Homework: *Do It the Chefs' Way*	5
6	Discussion: *Do It the Chefs' Way*	35
	Reflection: First 5 days of class	5

	1-2-3-4 Puzzle	30
	POW 2: Checkerboard Squares	15
	Homework: *Uncertain Answers* and *You're the Chef*	5
7	Discussion: *Uncertain Answers* and *You're the Chef*	20

Investigations

	Consecutive Sums	20
	Add It Up	30
	Consecutive Sums (continued)	15
	Homework: *Group Reflection*	5
8	*Consecutive Sums* (continued)	50
	Pattern Block Investigations	35
	Homework: *That's Odd!* and *Degree Discovery*	5
9	Discussion: *That's Odd!*	10
	Pattern Block Investigations (continued)	10
	Discussion: *Degree Discovery*	15
	Polygon Angles	30
	Discussion: *Group Reflection*	15
	Discussion: *POW 2: Checkerboard Squares*	10
	Homework: *An Angle Summary*	0
10	Discussion: *An Angle Summary*	10
	Presentations: *POW 2: Checkerboard Squares*	20

Putting It Together

	Squares and Scoops	20
	Diagonally Speaking	40
	Homework: *Another In-Outer*	0
11	Discussion: *Another In-Outer*	30
	Diagonally Speaking (continued)	15
	The Garden Border	25
	Discussion: Introduce portfolios	15
	Homework: *Border Varieties*	5
12	Discussion: *Border Varieties,* including equivalent expressions and the distributive law	40
	Patterns Portfolio	35

	Homework: *Patterns* Portfolio (finish)	5
	Homework: *Take-Home Assessment*	10
13	*In-Class Assessment*	45
	Discussion: *In-Class Assessment* and *Take-Home Assessment*	30
	Unit Reflection	15

Materials and Supplies

All IMP classrooms should have a set of standard supplies and equipment. Students are expected to have materials available for working at home on assignments and at school for classroom work. Lists of these standard supplies are included in the section "Materials and Supplies for the IMP Classroom" in *A Guide to IMP*. There is also a comprehensive list of materials for all units in Year 1.

Listed below are the supplies needed for this unit. General and activity-specific blackline masters are available for presentations on the overhead projector or for student worksheets. The masters are found in the *Patterns* Unit Resources.

Patterns

- Four tubs of pattern blocks

- A set of overhead pattern blocks

- A bag of beans or similar manipulative to serve as counters

- Cubes of two different colors to represent positive and negative numbers

- Sets of *Lonesome Llama* cards (one set for each group; included in the *Patterns* Blackline Masters)

More About Supplies

- Graph paper is a standard supply for IMP classrooms. Blackline masters of 1-Centimeter Graph Paper, ¼-Inch Graph Paper, and 1-Inch Graph Paper are provided so that you can make copies and transparencies for your classroom. (You'll find links to these masters in "Materials and Supplies for Year 1" of the Year 1 guide and in the Unit Resources for each unit.)

Assessing Progress

Patterns concludes with two formal unit assessments. In addition, there are many opportunities for more informal, ongoing assessment throughout the unit. For more information about assessment and grading, including general information about the end-of-unit assessments and how to use them, see "Assessment and Grading" in *A Guide to IMP*.

End-of-Unit Assessments

Each unit concludes with in-class and take-home assessments. The in-class assessment is intentionally short so that time pressures will not affect student performance. Students may use graphing calculators and their notes from previous work when they take the assessments. You can download unit assessments from the Patterns Unit Resources (or you can find them in the Blackline Masters).

Ongoing Assessments

Assessment is a component in providing the best possible ongoing instructional program for students. Ongoing assessment includes the daily work of determining how well students understand key ideas and what level of achievement they have attained in acquiring key skills.

Students' written and oral work provides many opportunities for teachers to gather this information. Here are some recommendations of written assignments and oral presentations to monitor especially carefully that will offer insight into student progress.

- Presentations on *Calculator Exploration:* These presentations will give you information on how comfortable students are with calculators and open-ended investigation.

- *Pulling Out Rules:* This activity will help you gauge how well students understand the basic ideas of In-Out tables and evaluate their ability to write rules to describe tables.

- *You're the Chef:* This summary activity will tell you how well students understand the arithmetic of positive and negative integers.

- Presentations on *Consecutive Sums:* These presentations will indicate how students are developing the ability to conduct independent mathematical investigations.

- *An Angular Summary:* This activity will help you gauge students' understanding of the sum of the angles in a polygon and related formulas.

- *Border Varieties:* This activity will reflect students' understanding of the use of variables.

Supplemental Activities

Patterns contains a variety of activities at the end of the student pages that you can use to supplement the regular unit material. These activities fall roughly into two categories.

- **Reinforcements** increase students' understanding of and comfort with concepts, techniques, and methods that are discussed in class and are central to the unit.

- **Extensions** allow students to explore ideas beyond those presented in the unit, including generalizations and abstractions of ideas.

The supplemental activities are presented in the teacher's guide and the student book in the approximate sequence in which you might use them. Below are specific recommendations about how each activity might work within the unit. You may wish to use some of these activities, especially the later ones, after the unit is completed.

Keep It Going (reinforcement) Students use patterns to find the next few terms of four number sequences and then describe the patterns they found.

The Number Magician (reinforcement) In essence, this activity describes a multistep In-Out machine. Students determine the original number that produces one particular answer and analyze the method used to determine the original number so quickly.

Whose Dog Is That? (extension) This logic puzzle, much like *Who's Who?*, gives students another opportunity to use organized thinking and to write clear explanations. Students are given several interlocking conditions and must use logical reasoning to determine a set of conclusions. Give students several days to work on the activity and to write up their results.

A Fractional Life (reinforcement) This activity is part of *The Greek Anthology,* a group of problems collected by ancient Greek mathematicians. It will help reinforce students' work with In-Out tables and can be used any time after In-Out tables are introduced.

Counting Llama Houses (extension) Students identify the ways in which the houses in *Lonesome Llama* differed and then determine how many different houses could have been created using these variations.

It's All Gone (reinforcement) In a variation on *Marcella's Bagels,* a man goes from store to store getting and spending money, winding up with no money in the end. Students are asked to determine how much money he had when he started.

1-2-3-4 Varieties (reinforcement) This activity adds a rule to those used in *1-2-3-4 Puzzle:* now the digits must appear in numeric order. In addition to finding expressions for the first 25 whole numbers, students are asked to find the greatest possible answer given these rules and to make up their own variations to the original activity.

Positive and Negative Ideas (extension) This activity extends students' work with hot and cold cubes. It asks them to consider other ways they might model integer arithmetic.

Chef Divisions (extension) This activity extends ideas introduced through the "hot-and-cold-cubes" model. While modeling division with hot and cold cubes, students think more deeply about the model and the reasoning involved in working with negative numbers.

More Broken Eggs (extension) In *The Broken Eggs,* students found a possible number of eggs the farmer might have had when her cart was knocked over. The task now is to look for other solutions, to find and describe a pattern for obtaining all the solutions, and to explain why all the solutions fit that pattern.

Three in a Row (extension) Students explore sums of three consecutive numbers as well as sums of other lengths. The activity is appropriate following the discussion of *That's Odd!*

Any Old Sum (extension) In this variation on the *Consecutive Sums* investigation, students examine sums that are not consecutive. In addition to extending ideas in *Consecutive Sums,* this activity gives students more experience with open-ended problems.

Getting Involved (reinforcement) Several activities in this unit—such as Role Reflections and Group Reflection—ask students to reflect on the process of working in groups. In this related activity, students are asked to reflect on a situation in which one person in a group is not contributing.

The General Theory of Consecutive Sums (extension) Students explore consecutive sums of integers. You may want to allow students several days to work on this challenging activity.

Infinite Proof (extension) Students are asked to prove that the square of every odd number is odd and that every prime number greater than 10 must have 1, 3, 7, or 9 as its units digit. The activity gives students the opportunity to see that proofs are possible in situations involving infinitely many cases.

Different Kinds of Checkerboards (extension) In this follow-up to *POW 2: Checkerboard Squares,* students find the number of squares on nonsquare checkerboards and search for a general rule for checkerboards of dimensions m by n.

Lots of Squares (extension) In this substantial investigation, students are asked to divide a square into different numbers of smaller squares. The goal is to determine which numbers of smaller squares are impossible and which are possible, and to prove their results. Assign the activity after students have worked on developing proofs.

A Protracted Engagement (reinforcement) In this open-ended activity, students are asked to decode a message created using angles of different sizes to correspond to different letters of the alphabet, and then to code a message of their own.

A Proof Gone Bad (reinforcement) Students are asked to explain the contradictions in another student's proof. Assign the activity after students have worked on developing proofs.

From Another Angle (extension) This activity extends students' work with pattern blocks and generalizes ideas in *An Angular Summary*.

From One to N (extension) The task in this activity, which is a natural outgrowth of the ideas in *Squares and Scoops,* is to find a simple expression in terms of *n* that allows one to find a sum without repeated addition. If students find such an expression, they look for a proof that their answer is correct.

Diagonals Illuminated (extension) This follow-up activity to *Diagonally Speaking* draws the distinction between recursive and closed-form rules and asks students to develop a closed-form rule for the number of diagonals of any polygon. Students are then asked to explain why this rule makes sense.

More About Borders (extension) This activity contains variations on the *Border Varieties* activity.

Programming Borders (extension) Building on ideas in the supplemental activity *More About Borders,* this activity asks students to write a program that answers some or all of the questions posed in *More About Borders.*

Patterns: A Mathematical Commentary

Eric Robinson, Ithaca College, Ithaca, New York

The goal of this commentary is to situate the mathematics in the unit *Patterns: Functions, Conjectures, and Proof* within the larger context of mathematics as a discipline. In my opinion, the material in this unit lies within the very *core* of modern mathematics. (What a perfect way to begin the curriculum program!)

Of course, part of the reason this material is placed within the mathematical core is because the unit deals with fundamental mathematical concepts: numbers, shapes (polygons), and functions. (I will mention more about functions later in the commentary.) However, as a mathematician, I want to focus on what gets me really excited about this unit: the environment it provides for students to engage in mathematical thinking and reasoning. And while mathematics has many facets, at its *center* lies mathematical thinking and reasoning: a methodology that includes exploring (investigating), conjecturing, explaining, and justifying certain phenomena. It is from mathematical thinking and reasoning activity that the body of mathematical facts emanates and grows—often building upon itself—and that mathematical algorithms, procedures, and structures are created.

Moreover, I think one of biggest motivations—if not the biggest—for most mathematicians to devote high energy to mathematics (and what obsesses them in particular instances—sometimes for years) is the chance to engage in mathematical inquiry: the search for answers to questions such as (1) What is going on here (mathematically)? (2) Why is it happening? (3) Can I be certain this always will happen? (4) Under what conditions will it happen this way? (That is, is this generalizable or extendable?) The first of these questions usually leads to a conjecture. The second stimulates a search for insight. The third motivates a search for validation (proof). Answers to the fourth extend the generality and applicability of a known result. Mathematicians do not have the luxury of knowing whether a conjecture is true or false before trying to prove it. They often look for evidence by exploring multiple examples and they probably will try a range of reasoning strategies in pursuit of their goal to confirm or reject the conjecture.

The point is that mathematicians' work centers on asking and answering questions. Now, if mathematicians answer questions, a mathematics textbook should be filled with rich questions—similar to those mentioned above. The *Patterns* unit (and, for that matter, the whole curriculum) is.

It is very heartening in this unit to see students engage in the search for mathematical insight, including the search for mathematical patterns, the formation of mathematical conjectures, verification or refutation of these conjectures, looking for generalizations and extensions of results, and abstracting properties from particular instances or examples. This is what mathematicians do. The reader easily can find instances of each of the aforementioned mathematical activities in multiple places within the *Patterns* unit. Now, the term "proof" commonly refers to (logical) justification of a result. So, narrowly construed, the word does not capture the rich mathematical reasoning aspects described above that accompany the search for proof. Fortunately, the *Patterns* unit begins to capture these mathematical reasoning characteristics.

Mathematics has been described as a "science of patterns." At the very least, this description is intended to emphasize the prominent place of patterns within the discipline. Indeed, patterns play multiple roles in mathematics. Here, we mention a few of theses roles. First, the search for patterns is a fundamental approach to creating conjectures. There are many open conjectures in current mathematics that relate to patterns. One open example (as of this writing) that is relatively easy to explain has to do with twin primes. Now, all prime numbers except the number 2 are odd. So, two different prime numbers greater than 2 must be at least two units apart. Some pairs of primes are exactly two units apart. For instance, 3 and 5 are two units apart; so are 5 and 7, 11 and 13, and 17 and 19. Two prime numbers (greater than 2) that are exactly two units apart are called twin primes. Many pairs of twin primes have been found, some quite a way along in the natural numbers, but seemingly with increasing scarcity. In the latter half of the 19th century it was conjectured that the pattern of existence of twin primes is unending. That is, there are infinitely many pairs of twin primes. But no one has been able to prove it—yet. (Twin primes are mentioned later in the IMP curriculum.) Study of twin primes has led to the discovery of many interesting patterns and results related to these numbers! (http://mathworld.wolfram.com/TwinPrimes.html)

In the *Patterns* unit under consideration, the activity *Consecutive Sums* is a good example of the role of patterns in developing conjectures. The request for explanation in that activity and the associated activity *That's Odd!* helps bring home the idea that finding patterns is only the first part of the mathematical inquisition; there is eventually a need to verify that the pattern persists and, hopefully, one will be able to answer why and how the pattern works.

Patterns are studied in many mathematical domains. They occur in the domains of numbers, shapes, and other mathematical structures. Sometimes patterns develop when they aren't expected and mathematicians can get excited when they see them. Michael Barnsley's "Chaos Game" provides this experience. (http://www.jgiesen.de/ChaosSpiel/ChaosEnglish.html) When the game is "played," a highly patterned deterministic fractal emerges as the *limit* of a random process.

While a pattern can serve as input for conjectures, a pattern can often be interpreted as the output of many a theorem. Almost any theorem can be thought of as a statement about patterns. The theorem in Euclidean geometry stating that the base angles of any isosceles triangle are always equal in measure is a statement of a persistent pattern in isosceles triangles.

How much data is necessary to make a conjecture about a pattern? Sometimes a little and sometimes a lot. To an experienced mathematician, sometimes one carefully constructed example can suggest a generalized pattern worth pursuing. On the other hand, sometimes conjectures with too little data turn out to be false. The great Pierre de Fermat (1601–1665) computed the numbers of the form $2^{2^n} + 1$ for $n = 0, 1, 2, 3,$ and 4. The results were 3, 5, 17, 257, and 65,537—all prime numbers. Fermat confidently concluded that all numbers of the form $2^{2^n} + 1$ are prime, where n is a non-negative integer. Fermat was wrong in this case. In fact, the very next case, $2^{2^5} + 1 = 4,294,967,297$, is not a prime number. It was laborious to check primality in numbers this large by hand in Fermat's day and Fermat obviously did not check. In 1732, Leonard Euler reasoned correctly that 641 is a factor of $2^{2^5} + 1$.

So, in searching for patterns, it is important to note that some data can suggest a definitive pattern and be correct and some conjectured results from the data about a pattern can fail (as happened for Fermat). It is also important to note that some data can suggest multiple possible patterns. This latter fact is nicely illustrated in the IMP activities *Inside Out* and *Pulling Out Rules*. Activities like *Pulling Out Rules* also hint at the general mathematical notion of curve fitting: finding functions that "fit" data sets. (Where the term *fit* is further defined—often in relation to the kinds of data sets or functions involved.)

Now, we have segued into the third major word in the title of the unit: functions. The unit treats functions as rules of assignment; as inputs paired with outputs. As such, functions can be described verbally, suggested in table form, represented graphically, or coded symbolically. Each representation has its advantages and disadvantages. The first two types of representations are encountered early in the unit. But, as the unit progresses, the foundation and motivation to express rules symbolically is clearly laid. Of course, this is one of the fundamental characteristics of algebraic thinking: using symbols (variables) to code a general relationship.

Functions appear in nearly all major branches of mathematics and they serve a variety of roles. For example, they may help us to represent change (as in differential calculus) or what appears to be the opposite: invariance (as elements of transformation groups used to describe symmetry patterns of geometric objects).

Finally, as illustrated by the study of symmetry groups just mentioned, patterns, functions, conjectures, and proof are not isolated objects, but rather are part of the interwoven fabric of mathematics. Indeed, this interconnectivity is yet another central aspect of modern mathematics. This interconnectedness is well illustrated in the *Patterns* unit and, for that matter, in the curriculum as a whole.

The Importance of Patterns

Intent

The activities in *The Importance of Patterns* introduce a crucial mathematical idea—functions and their representations—that will weave its way through the entire curriculum. Students begin to build their ability to tackle novel mathematical problems, a way of doing mathematics that permeates IMP. In addition, these activities begin to establish expectations for students' classroom interactions—as a whole class, in small groups, and individually—and written work throughout the course. Finally, students have their first experiences using graphing calculators.

Mathematics

The central mathematical idea in *The Importance of Patterns* is the concept of function, one of the fundamental unifying principles in mathematics. Functions are introduced using numeric and nonnumeric examples, with an emphasis on looking for patterns and describing those patterns verbally. The term **function** is introduced in the context of describing the *Out* as "a function of" the *In*. The discussion introduces the principle that a function cannot have more than one output for a given input as well as the concepts of the **domain** and **range** of a function. Students use **variables** and *algebraic expressions* to describe numeric functions. They also apply *In-Out tables* to mathematical problems and see the distinction between tables arising from context and tables that are simply collections of number pairs. In-Out tables are a standard method for representing functions and are central to this unit and the curriculum.

Variables play a vital role throughout mathematics, and they represent a major step toward mathematical abstraction. One major goal in *Patterns* is for students to use the symbolic language of algebra as shorthand to describe patterns, particularly arithmetic patterns in In-Out tables. Students use both their verbal description of a table's rule and the pattern of arithmetic for finding specific outputs in order to develop an algebraic expression that describes the rule. As part of this work, they begin to use the terms *variable* and *algebraic expression*. The concept of **equivalent expressions** is introduced in the context of seeing that different expressions give the same results for an In-Out table.

Progression

What's Next?

Past Experiences

POW 1: The Broken Eggs

Who's Who?

The Standard POW Write-up

Inside Out

Calculator Exploration

Pulling Out Rules

Lonesome Llama

Role Reflections

What's Next?

Intent

This first activity in the unit engages students in a series of questions for which there may be no familiar procedure or algorithm and for which there might be many solutions. It is students' first opportunity to do mathematics together.

Mathematics

This activity introduces the mathematical idea of a **sequence**. Students are asked to find patterns that fit a given sequence and then to use these patterns to predict the next few terms of the sequence. The search for patterns is a recurring theme in this unit and throughout the IMP curriculum. These early activities also build a foundation for the concept of **function**, one of the truly big ideas of algebra.

Progression

After you have introduced IMP to students as a somewhat different kind of textbook, and have made students aware that their classroom working environment will have certain characteristics (see "Characteristics of the IMP Classroom" in the Overview to the Interactive Mathematics Program), this activity will be their first experience of IMP and the "IMP classroom." Students will work on this activity in a small group of peers. They will be encouraged to be creative in describing their patterns and to share ideas with group members. The activity concludes with students discussing some of the patterns they identified.

Approximate Time

30 minutes

Classroom Organization

Small groups, followed by whole-class discussion

Doing the Activity

You might begin by asking one or more volunteers to read the instructions aloud. Emphasize that students are to do more than simply find the next few terms of each sequence; they are also to give a description of the pattern they see.

Offer some ideas about expectations for group collaboration and interaction—namely, that groups do both. Suggest a few methods to do so, such as occasionally asking what a neighbor found to compare to your result or what someone sees when you haven't been able to notice a pattern. Emphasize that all students are expected to ask for help when stuck and to help others when asked.

Have groups begin writing down their ideas. Students might work individually on the first two questions and then discuss those in their groups before moving on to the next pair of questions. Stopping to share ideas lets the groups hear what everyone is thinking and see that there is more than one possible pattern or approach. Sharing also helps students learn to value each other's thinking and to collaborate.

Circulate as students work and listen in on the discussions, limiting interventions in order to encourage students to rely on their own thinking and to work collaboratively within their groups. You might ask groups that need help some probing questions, such as those below.

In your own words, what is the pattern you found?

Does the pattern fit the terms of the sequence?

Does someone else have another way to describe this pattern?

Did you find other patterns that start the same way?

Can there be more than one correct pattern?

Remind students, if needed, to look for more than one possible pattern for each sequence or more than one way to describe a given pattern. If many groups seem stuck, you might interrupt and have a class discussion of the first question or two to clarify what is being asked.

When the majority of groups have finished Questions 1 through 6, you might bring the class together for discussion. As time allows, you can then have them turn to Questions 7 and 8.

Discussing and Debriefing the Activity

On the first day of the unit, it is important to establish a classroom climate where student thinking is valued and where it is safe to take the risk of sharing ideas.

Ask presenters to describe the patterns they found and the next terms these patterns led to. These descriptions can be very informal. Then ask for comments from the class. Have the class work together to make the pattern descriptions clear. (Over the course of this unit, you will ask students to make more precise statements, including algebraic descriptions of some patterns.)

It is important that students see different ways to describe a given pattern as well as different patterns that fit a given initial sequence. Ask if anyone found other ways to describe a given pattern. What other ways did you find to describe this pattern? For example, some students may describe the sequence in Question 4 (1, 2, 4, . . .) as "Double each term to get the next term," while others may say "Add each term to itself to get the next term." Although these two descriptions are different, they lead to the same continuation of the pattern, with the next three terms being 8, 16, and 32.

Did you find other patterns that start the same way? Some students may have discovered different patterns that fit the same opening terms of a given sequence. For instance, in Question 4 (1, 2, 4, . . .), the pattern could be "Add 1, then add 2, then add 3, and so on." In this pattern, the next three terms would be 7, 11, and 16 (rather than the 8, 16, and 32 for a doubling pattern).

Question 5 also offers more than one option. For example, the sequence could repeat the opening terms, 1, 3, 5, 7, 5, 3, over and over again (so it goes 1, 3, 5, 7, 5, 3, 1, 3, 5, 7, 5, 3, 1, 3, . . .), or it could follow 1, 3, 5, 7, 5, 3 with 1, 3, 5, 7, 9, 7, 5, 3 and then 1, 3, 5, 7, 9, 11, 9, 7, 5, 3, and so on.

When students present more than one description of a pattern or more than one pattern, ask if both can be right. Bring out the idea that any description or pattern that fits the opening terms of a sequence is as correct as any other.

As time allows, ask students to present sequences that their groups created for Questions 7 and 8. You might display these on the board or have groups try to figure out each other's patterns.

Key Questions

In your own words, what is the pattern you found?

Does the pattern fit the terms of the sequence?

Does someone else have another way to describe this pattern?

Did you find other patterns that start the same way?

Can there be more than one correct pattern?

Supplemental Activity

Keep It Going (reinforcement) asks students to use patterns to find the next few terms of four number sequences and to describe the patterns they found.

Past Experiences

Intent

This individual activity is best used as homework, the first assignment of the school year. Including a writing assignment like this one will establish several expectations for the course.

Purposeful homework will be assigned every day, and students' work on these assignments will be an important part of the course.

Students will be asked to put their thinking—about mathematics and about themselves as learners of mathematics—to paper.

All students' thoughts and ideas about the mathematics they are learning are crucial to the success of the course.

Successful collaboration to do and learn mathematics is a key feature of this course.

Mathematics

At first glance, this assignment does not look particularly mathematical. However, a growing body of research suggests that successful mathematical problem solvers are reflective thinkers. They know mathematics, and they know about mathematics as a discipline. They are aware of themselves as mathematics learners, and they can think about their own thinking—monitoring progress, evaluating strategies, choosing among skills and tools—while doing mathematics. Psychologists call this *metacognition*, and it is a hallmark of the thinking of effective problem solvers. In this activity, students are asked—perhaps for the first time (and certainly not the last time in this program)—to reflect on some of their experiences as mathematics students.

Progression

This activity is designed to be done as homework after the first class and to be discussed, in small groups and as a whole group, in the next class.

Approximate Time

10 minutes for introduction

20 minutes for activity (at home or in class)

10 minutes for discussion

Classroom Organization

Whole class, then individuals, followed by small groups

Doing the Activity

Take the time to share your expectations for this assignment and homework in general, including what you expect from students and what students can do if they don't understand an assignment. Telling students that you want to learn more about them and their backgrounds, and that you will not be grading their essays, but just recording whether they completed the assignment, may encourage them to do the assignment and to share honestly. One important goal of the first few homework assignments is to help students establish a pattern of doing their homework regularly.

Also impress upon students that they need to save their work throughout the unit, as they will be asked to include their written work on this assignment and others in the portfolios they will create at the end of this unit.

For the next day's discussion, you might want students to share their essays in their groups. If you plan to follow this suggestion, let students know now that other students will be reading their written work.

Discussing and Debriefing the Activity

Students can read the essays of the other members of their groups. You might suggest that after reading each other's thoughts and experiences, students answer the Key Questions listed below, perhaps displaying these or similar discussion questions on a transparency. Then students can share with the class the themes their groups encountered.

This is a good opportunity to reiterate that class participation—written, oral, and physical; in groups, individually, and with the whole class—is essential for success.

Key Questions

What are some of the important mathematical ideas you have studied?

How are your group's ideas about your most and least helpful learning experiences similar? How are they different?

How are your experiences, thoughts, and feelings about working with others similar? How are they different?

POW 1: The Broken Eggs

Intent

As the first POW, or Problem of the Week, *The Broken Eggs* is students' first opportunity to work on a substantial problem over several days and communicate the results of their work in writing, using a format that will carry across the four years of the program. (See "Problems of the Week" in the Overview to the Interactive Mathematics Program.)

Mathematics

This POW is a version of a well-known problem in number theory. Here is a translation from a seventh-century text written by the Hindu mathematician Brahmagupta: *An old woman goes to market, and a horse steps on her basket and crushes the eggs. The rider offers to pay for the damages and asks her how many eggs she had brought. She does not remember the exact number, but when she had taken them out two at a time, there was one egg left. The same happened when she picked them out three, four, five, and six at a time, but when she took them out seven at a time they came out even. What is the smallest number of eggs she could have had?* A similar problem was posed by the Chinese scholar Sun Tsu Suan-Ching in the third century: *There are certain things whose number is unknown. Repeatedly divided by 3, the remainder is 2; by 5 the remainder is 3; and by 7 the remainder is 2. What will be the number of things?*

In the activity, students search for numbers divisible by 7, but when divided by each of numbers 2 through 6 leave a remainder of 1. To find solutions to this problem, students must examine multiples of 7 and remainders when dividing by 2 through 6, and reason about patterns in these results. *The Broken Eggs* problem has many solutions, creating a complex task that will allow any high school student to begin to work on the question and all to pursue it as far as their interest (and time) allows (see "About Solutions to Activities" in the Overview to the Interactive Mathematics Program).

Progression

Students will work on this POW primarily outside of class. This unit is carefully designed to support student success, especially with this first long-term, problem-solving and writing project. The problem is posed early in *The Importance of Patterns* and revisited at several points over the next few class meetings. Three students will present their solutions to the class, and all will turn in their written work.

Approximate Time

5 minutes for introduction

30 minutes for groups to begin exploration

20 minutes for introducing POW write-ups

30 minutes for individuals (homework; do and write Process)

5 minutes for small-group discussion

30 minutes for individuals (homework; complete Process)

15 minutes for discussion findings and of next phases of write-up

15 minutes for discussion of presentation expectations

30 minutes for individuals (homework; complete POW write-up)

15 minutes for presentations

Classroom Organization

Individuals and small groups, concluding with whole-class presentations and class discussion

Materials

Presenters will need presentation materials, such as transparencies and pens, a few days prior to the due date.

Doing the Activity

Please be thoughtful about the extended timeframe over which the activity occurs. The more support you offer students in this, their first long-term and significant problem-solving and writing activity, the more successful they will be in the subsequent writing activities they will encounter.

The commentary below divvies the focus of this supportive work into several "phases." This is not meant to indicate sequential days; depending on your scheduling situation, consider these the segments necessary to support student work, however you parse it into your daily plans.

Introduce the Problem

You might begin this activity by having one or more volunteers read it aloud, through the section Your Task. Tell students that they will be working on this problem in groups, sharing ideas and insights. They should keep notes on how they and their groups work on the problem, as they will be discussing these things in their final write-ups.

Begin Exploring

Have students work on the problem in groups. Although some students might find a solution during this initial work, this is not an expectation for the first day.

If groups are stuck, ask what they have tried. It is important that students have time to reach conclusions at their own pace. To clarify whether they understand the problem, you might ask why 49 is not a correct answer. You might also suggest they consider this simplified problem: Suppose the farmer remembered that only when she put the eggs in groups of either two or five, there was one egg left over. What would be some possibilities for the number of eggs in that situation?

If any groups find the answer 301 today, you can urge them to look for other solutions. If groups need further challenges, they can look for a general solution or a description of how to find other solutions, and then for an explanation of how they know their general solution includes all possibilities.

The Standard POW Write-Up

Have students read *The Standard POW Write-Up.* Most POWs will have a Write-up section that uses the basic components listed in *The Standard POW Write-Up.* The write-up instructions will often simply refer to these components by name, giving additional details only when the write-up differs from the basic model provided here. You may want to post the five write-up components on the wall.

Have students work individually for a while to create problem statements for the POW. Then have them share their ideas in their groups, and have each group use these ideas to create the best problem statement possible.

Ask one or two groups to share their problem statements with the class. In the discussion, bring out that the problem statement should not simply repeat the problem as originally stated, but should try to focus on the essentials. You might work with students to distinguish between the "story" aspects of the problem and its mathematical core. With this first POW, students might include both aspects in their problem statements, but over the course of the curriculum, they should gradually move toward an emphasis on the mathematical elements of the POWs.

You might remind students that taking notes as they work will help with the Process part of their write-ups. Encourage them to collaborate with classmates. You might mention explicitly that you do not consider it cheating to work with someone else on a homework assignment or POW, as long as students acknowledge that collaboration in their write-ups. On the other hand, students should not simply copy each other's work or allow others to copy from theirs. You might also offer advice about how to help each other, such as by giving a hint or asking a leading question. Point out that if they give a friend the answer, they deprive the friend of much of the learning experience.

Begin Writing the Process

Two or three days after assigning this first POW, assign for homework simply working on the POW for 20 to 30 minutes. Tell students they should come to class tomorrow with a portion of their Process written. If they are using word-processing software for their write-ups, they should bring a printout of this draft.

Encourage students to keep notes about ideas they have and things they try and to pause occasionally to add to the Process portion of their write-ups. A structure such as work 7 minutes, write for 3 minutes might help students achieve this goal.

When students return to class, encourage them to share ideas in their groups. Then explicitly instruct them to share what they have done in writing the Process. It can be valuable to have students pass their write-ups around their groups to see how others are recording their solution methods.

Finish Writing the Process

On the very next day (the day the homework above is due), have students spend 20 to 30 minutes working on the problem and completing the write-up of the Process. Again, they should bring a draft to class.

In class, again encourage groups to share findings from their investigations the previous evening, including reading one another's Process, perhaps in pairs this time.

Follow this with a class discussion. Ask students to describe what they think is in the Process section, as they have written it over the past few days. Return to the *The Standard POW Write-Up* reference page, and ask students to compare their impressions with the description of the Process section here.

Remind students that they have three sections left to write: Solution, Extensions, and Self-assessment.

Completion and Presentation Preparation

Use the contexts of *Marcella's Bagels* and *Extended Bagels* to remind students of the write-up structures and expectations.

As this will be the first POW presentation, ask for three volunteers to present their work. You may find it easier to get volunteers if you mention that this first group of presenters will get some extra guidance. (For future POWs, either select students at random, choosing from among those who have not yet done POW presentations, until everyone has had a turn, or ask for volunteers, again explaining the expectation that everyone will present once before cycling through again.)

Discuss with the class what will be expected of presenters and of the audience. Emphasize that presentations are to be discussions about *ideas*. It is important that presenters prepare to share what they learned about the problem and not feel pressured to present "the" answer.

Audience members should listen to discover what presenters have figured out, how they approached the problem, and the reasoning behind their conclusions. The audience will be expected to ask clarifying questions, such as "I don't understand how you arrived at this conclusion; I seem to get ___" or "That idea seems to contradict ___."

Presenters are to use transparencies to *help with* their presentations, rather than to *be* their presentations. In other words, they will not present only what is on the transparencies, but should plan to explain the problem, using the transparencies to save the trouble of writing as they talk. They can include diagrams, numeric calculations, and whatever else might be helpful.

Also, presenters may need reminding to plan to talk about *all* parts of the write-up, not only about their solutions. (In later POW presentations, students may find certain sections need less attention, especially given only 5 minutes to present.) Finally, encourage them to make any writing on transparencies large enough to be readable.

You might meet briefly with volunteers to address any questions or concerns they have and to give them transparencies and pens for preparing their presentations.

Once presenters are selected and expectations for the presentations have been communicated, tell everyone that the homework is to finish the write-up.

Discussing and Debriefing the Activity

Presentations

Because making presentations is very difficult for some students, these first POW presenters might have a tough time. Because of their willingness to volunteer, they deserve special consideration from the audience and assistance from the teacher.

Briefly remind the audience of these expectations (which you may want to post):

Acknowledge the effort and courage of the presenter, regardless of the quality of the presentation.

Treat each other with respect and listen attentively.

Listen for what the presenter learned, ask questions when you don't understand, and challenge things you think are incorrect. Being respectful does not mean being passive. It is not disrespectful to question, add to, or challenge each other's work if it is done in the proper spirit.

Multiple routes to the solution: Encourage students to ask questions during the presentations. After all the presentations are over, ask if anyone has anything else to add. Be sure students realize that this invitation includes presenting a different method for finding or explaining an answer—they do not have to have a new or a different answer.

More than one solution: If the presentations did not deal with the issue of the POW having more than one answer, bring that up now. Many students may have stopped exploring when they found that 301 fits all the given conditions. An important mathematical question to ask is, Is this problem one of those that has more than one answer? This problem, like many others, does have more than one answer.

It is not necessary at this time that students find the general expression for all possible solutions, but they should recognize the possibility of multiple solutions. The supplemental activity *More Broken Eggs* asks students for the general solution.

Key Questions

How do you know that (301, for example) is a solution? Is not a solution?

Do you suspect there are other solutions? Why?

Supplemental Activity

More Broken Eggs (extension) expands on *The Broken Eggs*, in which students found a possible number of eggs the farmer might have had when her cart turned over. The task now is to look for other solutions, to find and describe a pattern for obtaining all the solutions, and to explain why all the solutions fit that pattern.

Who's Who?

Intent

This activity is included early in the unit to engage students in the important processes of logical reasoning and proof.

Mathematics

This activity presents interlocking sets of conditions. Using these conditions, students must identify "who's who" and are asked to provide a convincing argument—a proof—of their conclusion. Issues of proof arise repeatedly throughout the curriculum and in daily interactions. Most significantly, students are always expected to justify their solutions, to convince others, and to be convinced by others.

The reasoning students will use to analyze the stated conditions, to make conjectures about the solution, to test those conjectures to convince themselves that their solution meets the stated conditions, and then to determine whether their solution is unique—that is, to prove their solution—is at the heart of what it means to do mathematics.

Progression

Students are asked to find a solution to this puzzle and to determine whether that solution is unique. The activity also gives students the chance to use two components of POW write-ups: Process and Solution.

Approximate Time

5 minutes for introduction

20 minutes for activity (at home or in class)

15 minutes for discussion

Classroom Organization

Individuals, followed by whole-class discussion

Doing the Activity

Review what is expected in the activity. Urge students to start taking notes as soon as they begin thinking about the problem and to use those notes in the Process portion of their write-ups. Emphasize that the Solution part of their write-ups must demonstrate how they are certain of their solution and especially how they know that it is the only solution. (Some students might be unfamiliar with the game of Hearts, but they don't have to know anything about cards or this card game to do this activity.)

Discussing and Debriefing the Activity

Some students might approach this activity using logic. For example, because Felicia passes her cards to the ninth grader, she can't be the person who passes to the eleventh grader.

Others might list all possible cases. With only three students, it is not too difficult to list all possible grade levels and all possible arrangements of the students around the table, and then see which one meets all the conditions.

Whatever approach they use, students have the opportunity to engage in clear, logical argument to explain why their solution is unique.

Give students a few minutes in their groups to compare how they answered Question 1. Suggest that they focus not only on the answer, but also on how they worked on the problem—that is, on the Process component of the write-up.

As you circulate and listen in on the discussions, identify students who approached various parts of the activity in interesting ways, and ask them to present those parts to the class. Try to get several methods presented, both to describe the approaches themselves and to emphasize that there are many possibilities.

When students are convinced that there is a single solution, raise the issue of whether there are other answers, and ask students to explain how they can be sure there is only one. Is this one of those problems that has more than one answer? As with the discussion of solution methods, encourage different approaches.

After several students have offered explanations of why the answer is unique, ask whether students are completely convinced by these arguments. How convinced are you? Use this opportunity to review the word **proof**. Clarify that in this problem, a complete proof involves two aspects:

Showing that the particular seating arrangement satisfies the conditions in the problem; that is, that a solution *exists*

Showing that no other seating arrangement fits those conditions; that is, that the solution is *unique.*

Key Questions

Is this one of those problems that has more than one answer?

How convinced are you?

Supplemental Activities

The Number Magician (reinforcement) asks students to determine the original number that produces one particular answer and to analyze the method used to determine the original number so quickly.

Whose Dog Is That? (extension) is a logic puzzle much like *Who's Who?* Students are given several interlocking conditions and must use logical reasoning to determine a set of conclusions.

The Standard POW Write-Up

Intent

This reference page introduces students to Problems of the Week and to the standard POW write-up. Students will also refer to this reference page throughout the year to aid with their POW write-ups.

Mathematics

Communicating about mathematical thinking is an important part of doing mathematics. This reference page is designed to support students' written communication about their findings when exploring large mathematical problems. Stating the problem, discussing one's methods, and concluding succinctly and with justification, so that a reader will understand what has been written, should be the goal of every student writer. By suggesting extensions to the problem, students will be saying that the mathematics has not been fully explored, given the time constraints. Their self-assessment will be an evaluation of the effort and quality of their work, what they take pride in, and what they wish they could have done better.

Progression

Initially, this reference page will support students as they assemble a paper that communicates the work they did and what they learned. As they become more comfortable with the process of writing about mathematics, the page should be returned to and discussed occasionally. It will prompt students to think more deeply and to share ideas with other students about what each section of the POW write-up means, to evaluate each others' papers, and to focus on improving portions of their own work.

Approximate Time

20 minutes for discussion

Inside Out

Intent

This activity introduces students to a powerful representation of functions, In-Out tables, that will be used repeatedly throughout the IMP curriculum. The In-Out machine metaphor is used to introduce In-Out tables and the terms *input* and *output*. In their exploration, students experiment, make conjectures, and work toward getting clear verbal statements of the rules.

Mathematics

A powerful way to think about a **function** is as a machine. The thing put in is called the *input* (the *In*, for short); the thing that comes out is called the *output* (the *Out*). An *In-Out table* is one representation of a function. Other representations include graphs and symbolic rules. (Students will study these representations in *The Overland Trail*.) We often use the word *function* in such phrases as "the *Out* is a function of the *In*," which means that the *Out* value depends on, or is determined by, the *In* value.

Progression

This activity works well as homework. Students are introduced to the idea of an In-Out machine and learn to keep track of what it can do by using an In-Out table. They attempt to decode some In-Out tables and then create two of their own.

Approximate Time

15 minutes for introduction

20 minutes for activity (at home or in class)

45 minutes for discussion

Classroom Organization

Whole-class introduction, followed by individuals, then small groups, concluding with whole-class discussion

Doing the Activity

Introduce the idea of an *In-Out machine.* Many students may have been exposed to this method of representing functions, but others may not have been. Explain that something, often a number, is put into the machine. The machine does something to that object, and something comes out. The thing put in is called the *input* (the *In*, for short); the thing that comes out is the *output* (the *Out*). With your first example or two, you might show inputs as if they were actually being put into a machine and outputs as if they were coming out of the machine.

A nice way to start is to have students try to figure out what the machine is doing without any initial information. For example, you can ask, What happens if I put in the number 5? Have students make guesses, though they will probably recognize that they can't possibly know for sure. Offer clues, such as "too high" or "too low," until someone gets the answer you have in mind. You may want to have several options in mind at first so that it takes more than one guess.

Then ask for another number to use as the input and have students guess the corresponding output. Continue in this way, as students gradually get additional information about your mystery machine. At some point, ask, **How might you keep a record of the information?**

When most students seem to have figured out the rule, demonstrate how to organize the information using an In-Out table. For example, you might have a table like this for an "add 4 machine."

In	Out
5	9
3	7
0	4
9	?

Can you figure out a rule for this table? Ask students to tell you a rule in words. Work toward getting them to express rules completely, such as "The *Out* is 4 more than the *In*" or "You get the *Out* by adding 4 to the *In*" rather than just "Add 4." Write the rule next to the table.

Offer another example of an In-Out machine, and develop another table of student guesses. Name the input and the output parts of the table, using the metaphor of the machine, and name the table an In-Out table.

Next, give students an In-Out table, and ask them to guess the rule for the In-Out machine that the table is associated with. Again, write the rule as, "The *Out* is . . ."

Discussing and Debriefing the Activity

Ask students, in their groups, to share the rules they found for the In-Out tables in Questions 1 to 5. Also ask each group to prepare a transparency of one of the tables. Explain to the class that as they prepare to present, you would like them to focus more on their thinking process for identifying the rule than on the missing numbers. Also, presenters should write each rule as a complete sentence, beginning with "The *Out* is . . ."

During the presentations, encourage the audience to ask how the presenting group found each particular missing item or rule.

Questions 1 and 2: These two numeric tables represent what students will later classify as linear functions. Some students may have trouble finding the pattern in Question 2. Ask students who have found it to explain, if they can, the process they used to discover a rule to fit the information.

Question 3: The rule generally used for this table is that the *Out* is 1 less than the number of letters in the *In.* Based on this rule, you may want to bring out that there are many possible choices for the missing inputs, but only one choice for each missing output.

Question 4: This is probably the most challenging In-Out table in the activity, as it does not fit any standard idea of what constitutes mathematics, and because there is no simple algorithm for finding a relationship between the inputs and the outputs. As there is no explicit numeric information in the inputs, the first stage in thinking about this problem is to identify something in the pictures that can be associated with numbers. Students may have a variety of ideas about how to do this and will then need to find a rule that connects the numeric information in the pictures to the numbers as outputs.

If everyone is stuck on this problem, ask what changes from picture to picture and use this information to build a new table. For example, students might focus on the number of eyes, in which case they can see the table in Question 4 as equivalent to this table.

In	Out
1	3
3	11
4	15
?	7

Question 5: One rule that works is that the *Out* is the second vowel of the *In.* But students may have other ideas. Some may have decided, based on the first three rows, that the *Out* is the first vowel of the *In.* If so, use this opportunity to remind students to check their rules against all the given information.

Another possibility that may arise is that the *Out* alternates between the fourth and third letter of the *In.* That is, I is the fourth letter of *division*, E is the third letter of *ever*, O is the fourth letter of *opportunity*, and so on. If this or a similar suggestion arises, bring out that in this pattern, the *Out* depends on the sequence in which the input values occur rather than only on the value of the *In.* If the order of pairs shown in the table is changed, this pattern will no longer exist.

Make sure to discuss the "can't be done" entry for Question 5. Ask students what sense they made of it. They might respond with statements such as these.

To get "can't be done" as the *Out,* you need to put in a word with only one vowel.

You can only use words with at least two vowels for this table, so words with only one vowel can't be done.

Before introducing the term **function**, it is important to identify the distinction between functions and arbitrary tables of data. To illustrate, ask students what they think about a table like this one.

In	Out
3	5
2	8
3	7

Bring out that there is something unusual here, as there are two different outputs for the same input. In terms of the metaphor of an In-Out machine, you might identify this as a "broken machine."

With this as background, introduce *function* as the formal mathematical term that roughly corresponds to the idea of a "working" In-Out table. You might also use the phrase *function machine* as another term for an In-Out machine. The key idea is that a function must be consistent. That is, it must give the same output every time a particular input is used.

A related idea is that the output should not depend on where a given input is listed in a table. So a rule such as "The Out alternates between the fourth and third letter of the In" (see the earlier discussion of Question 5) does not describe a function.

Also include a case in which different inputs have the same output, such as Question 5. In other words, bring out that functions can't produce different outputs for the same input, but they are allowed to produce the same outputs from different inputs.

Explain that the concept of a function is one of the major unifying ideas of mathematics and that students will be working with functions throughout their mathematics program. You might mention that rule, table, and function are often used almost interchangeably in informal mathematical work, even though the terms technically have different meanings.

Introduce the term **domain** for the set of things that are allowable as inputs for a given In-Out table. Ask, What things are allowable as inputs for each table in last night's homework? You can bring out that in Questions 1 and 2, the *In* must be a number, while in Questions 3 and 5, it must be a word (or perhaps any sequence of letters). In Question 4, the *In* should probably be a picture similar to those shown.

In their work with In-Out tables, students have used such rules as "The *Out* is twice the *In*" or "You get the *Out* by adding 5 to the *In*." In the context of a specific example, you can bring out that a table will generally show only an incomplete picture of a function. It may have enough information to strongly suggest how the rule works, though students will already have seen that this is open to interpretation. But even if we settle on one specific rule, we can't tell from the table exactly what the domain is. Again, we can make an assumption about this, but usually it is only a guess.

Most often, the domain is an infinite set, and thus the function consists of an infinite number of In-Out pairs. Students should recognize that the table can only display a few of these pairs. You might indicate that because of this, we say that the table represents the function, but that technically the function is more than what is shown in the table.

Introduce the term **range** for the set of things that can be outputs for a given In-Out table. This set depends on what the domain is. For instance, for the doubling rule, if the domain is restricted to the whole numbers, then the range consists of the even whole numbers; but if the domain also includes positive fractions, then the range includes all whole numbers as well as all positive fractions.

Question 6: Have students exchange their In-Out tables and look for rules for the tables their fellow group members created. Each group can copy onto a sheet of poster paper two or three favorites from among the tables they created. They should make their tables big enough so that the entire class can read them when the poster is on the wall.

When groups have displayed their posters, they should attempt to find rules for the tables posted by other groups.

Ask the class whether there are specific examples they want to discuss or with which they had difficulty. You can have the group that created the problem or students from other groups offer hints on how to find a rule.

Key Questions

Did you see a method for finding the missing input?

What makes this problem difficult?

What do you think about this table?

Supplemental Activity

A Fractional Life (reinforcement) is part of *The Greek Anthology*, a group of problems collected by ancient Greek mathematicians.

Calculator Exploration

Intent

Graphing calculators will be part of students' tools for doing mathematics throughout the IMP program. This activity will give some students their first chance to learn how these calculators work. It also offers students an opportunity to make a short presentation of something they have discovered.

Mathematics

Given the versatility and power of the graphing calculator, and the wide variety of prior experiences students are likely to bring to this open-ended activity, students' explorations will probably range widely. However, there are some important mathematical issues students will encounter:

- How to handle order of operations on a calculator

  Calculators will evaluate such expressions as 17 − 6 ÷ 3 + 4 x 9^2, without the need for parentheses, by doing the exponent first, then multiplication and division, and finally addition and subtraction.

- How to use the built-in mathematical functions

  To find the square root of 3 you must access the square root function before the number 3, but to find 5! you must access the factorial function after the number 5.

  Sin(30) will not be 0.5 unless the mode is set to degrees.

- The graphing capabilities of the calculator

Progression

Students will explore their calculators in pairs and then share discoveries with the class.

Approximate Time

20 minutes for activity

20 minutes for discussion

Classroom Organization

Pairs, followed by whole-class discussion

Materials

Calculator guidebooks or manuals

TI Calculator Basics (optional), *IMP Year 1: Calculator Notes for the TI-83/84 Family of Calculators*

Overhead calculator

Doing the Activity

Tell students that they will be using a graphing calculator or handheld very often in their math class—so often that it will become a tool with which they think and explore, rather than simply calculate.

Have students read the activity on their own, and then highlight that they are to work with a partner to learn whatever they can about the calculator and, later, to demonstrate something they learned.

Give students time to work on this open-ended activity on their own, free from intervention. Through this experience, they may come to understand that they can learn about calculators, by trial and error, which will help them feel confident with these tools in the future.

Students should focus on simply learning how their calculator works. If you notice a pair fretting about not learning something in particular, encourage them with a reminder that the goal is to explore the calculator until they discover something new. You might also suggest that they explore a button that looks interesting to them. Or ask them to think of something they frequently do in math class and see if they can figure out how to do it on this calculator.

Encourage students to be thoughtful in their preparations to present. Assure them that their presentations can be simple, such as, "If you press this key, such-and-such happens." They don't necessarily have to learn how to accomplish something useful to make the information worth reporting.

Students may want to use manuals to learn how to do a specific activity or to find out what kinds of things the calculator can do. As you circulate, suggest to individual students that they prepare presentations on particular topics.

Discussing and Debriefing the Activity

Give pairs a short time to organize what they learned and to prepare their presentations. Pairs should probably be ready with several ideas to present, so they will have something available if another group presents one of their ideas.

Have pairs make their presentations in an appropriate sequence. For instance, schedule presentations that focus on more elementary aspects of calculator use before those on more advanced or obscure topics.

Students should present using an overhead calculator or appropriate software.

Pulling Out Rules

Intent

This activity gives students more opportunities to find and express rules for In-Out tables, both in words and symbolically, and to use an In-Out table as a problem-solving tool.

Mathematics

This activity raises several mathematical issues.

Finding rules for In-Out tables.

Developing symbol sense. For example, "The *Out* is 2 times the *In*, then add 3," Out = 2 In + 3, Out = 2 x + 3, and $y = 2x + 3$ are equivalent and increasingly abstract ways to use symbols to summarize the rule for Question 1a.

Confronting the idea that the number of data points in a table affects the number of rules that will explain all the data.

Using a function as a mathematical model of a quantitative situation, and then using the model to solve a problem related to that situation.

Introducing the terms **variable**, algebraic expression, **coefficient**, and constant **term.**

Progression

Students are first asked to find rules for In-Out tables that each contain four pairs of numbers. Then they are asked to generate many possible rules that fit tables with only one or two pairs of numbers. Finally, they are presented with a problem for which an In-Out table is a helpful solution tool.

Approximate Time

25 minutes for activity

20 minutes for discussion

Classroom Organization

Individuals, then groups, followed by whole-class discussion

Doing the Activity

Tell students that in this activity they will look for rules for more In-Out tables.

Discussing and Debriefing the Activity

Students should begin by asking questions of and comparing results with their group members. During this time, ask some groups to prepare presentations for one of the tables.

During presentations, encourage students to talk about the strategies they used to find their rules. Ensure that rules are presented as complete sentences, and record these sentences on the board.

Now that students have worked with a variety of In-Out tables, it may be valuable to let them share techniques that they have found for finding rules. As they do, help them to clarify their own thinking and encourage them to challenge each other to explain their ideas clearly.

Make sure a variety of rules for each table are shared. In Question 2, with only one data point to explain, students can find an infinite number of possible rules. If an In-Out table contains two points, as in Question 3, there are still infinitely many possible rules, but there is now only one linear rule that will work.

Question 4a asks students to create an In-Out table to analyze a real-life situation.

Number of volunteers	Number of bags of weeds pulled
1	2
2	5
3	8

Some students will answer Question 4b by extending the table using the pattern "Add 3 bags for each additional person" until they reach an output that is large enough. Some might find and employ a rule that fits the table *Number of bags* = 3 • (*Number of volunteers*) + 1.

Using either approach, students will discover that 30 does not appear as an *Out* if they stick to whole numbers for the *In*. **So what should the supervisor do?** To address this question, students must make sense of their work so far in light of the problem context. Their ideas could include these.

- The supervisor should "play it safe" by getting 11 volunteers, taking into account that the job might actually involve more than 30 bags of weeds and that some volunteers might work faster than others.

- The supervisor should get 10 volunteers and either make them work extra hard or join in the work as needed to finish the job.

There is no right answer; each of these suggestions (and other possible ideas) makes sense in this context.

Begin moving students from rules expressed in words to rules expressed symbolically. The discussion may be richer if this is a two-operation rule, such as "Get the Out by tripling the In and then adding 1."

Have students find the Out that goes with each of several In values for their rules. For the rule just stated, the table might look something like this.

In	Out
1	4
3	10
6	19
5	16

Help students record the rules as *algebraic expressions*. Students already know symbols for the numbers and operations involved, and most will have used symbols to replace unknown quantities. Remind them how much they know already about writing algebraic expressions.

With a few volunteered responses, students should arrive at an algebraic expression that all agree finishes the sentence. Record the expression in the In-Out table.

In	Out
1	4
3	10
6	19
5	16
t	$3 \bullet t + 1$

Below are a few additional ideas about algebraic notation and terminology that you can either elicit from students or simply present, using the context of the table just discussed.

- It is conventional to abbreviate $3 \bullet t + 1$ as $3t + 1$.

- It is acceptable to write $t \bullet 3$, unusual to write $t3$, and most common to write $3t$. Emphasize that omitting the multiplication sign is simply a convention of notation—that is, an agreement among mathematicians to write things a certain way. There is nothing inherently wrong about using a multiplication sign between a number and a variable or about placing the variable in front of the coefficient—it's just not usually done that way.

- The letter t is a **variable**. Rather than formally defining the term, you might just say that a variable is a letter that is being used to represent a general case.

- $3t + 1$ is an *algebraic expression*. A number used to multiply a variable, such as 3 in the expression $3t + 1$, is a **coefficient**.

- A number by itself that is added or subtracted in such an expression, such as 1 in the expression $3t + 1$, is a *constant term*.

After introducing this terminology and notation, you might ask students to express some of the other rules they found for the tables in Questions 2 and 3 in algebraic form. Using a variety of letters for the inputs emphasizes that the particular letter chosen has no significance.

Key Questions

What rules have we found for this table?

Are these rules really different? That is, would they lead to different tables? Or are they different ways of stating the same rule?

What should the supervisor do?

Lonesome Llama

Intent

This activity is designed to promote student cooperation and communication about mathematics by making the process of working together on a mathematical task an explicit learning focus.

Mathematics

Students will, within certain constraints, be trying to identify the unique card in a stack of 46 cards. The characteristics that distinguish the cards are mathematical—such as the number, type, and orientation of geometric figures—so students will be communicating about mathematics. The mathematical goals of this activity are for students to develop ways to describe the distinctive features within a set of diagrams that are largely alike and to develop a procedure for sorting the diagrams by those features.

Progression

After students, working in small groups, have found the singleton card or made sufficient progress, a whole-class discussion can focus on how they worked and what they discovered.

Approximate Time

30 minutes

Classroom Organization

Groups, followed by whole-class discussion

Materials

Lonesome Llama blackline master (46 cards for each group)

Doing the Activity

Understanding a bit about group dynamics can make a group a better team and enable students to get more out of the experience. The main purpose of *Lonesome Llama* is to get students to look at group processes and roles while they are engaged in problem solving. Everyone must participate in order to complete the task successfully.

Before passing out the sets of cards, have students read through the entire activity, and take some time to review the rules. Emphasize that students don't get to look at each other's cards until the activity is completed (that would make the activity way too easy!) and that what students say about how they work with each other is as important as what they learn about the cards.

Then hand out one set of cards, face down, to each group, and ask a student to deal out the cards approximately equally among the group members. (Because there are 46 cards per set, students in a given group won't all get the same number of cards.) Each student can then look at his or her own cards. Although students will have read the rules, you will likely need to review the rules one at a time, with students looking at their own cards, to ensure that everyone understands them.

Circulate as students work, ensuring that they follow the rules and observing how they are working together.

You might instruct the members of groups that finish early to respond in writing, privately, to this prompt: **What makes a group work well?** When they finish writing, you might have them begin the activity *Role Reflections*.

Discussing and Debriefing the Activity

Ask students for comments about the activity:

How did you know when you were done? How confident were you in knowing you had solved the problem? Why were you so confident?

What mathematics was involved in this activity? What else was mathematical about the ways your group worked?

This last question can be an opportunity to mention that mathematics involves not only knowing terms and facts, being able to use them efficiently and accurately, and solving problems; but also being able to reason, communicate, and share ideas with others so that you can do things as a team.

Ask for volunteers to share their ideas about the prompt, **What makes a group work well?** Students may choose to read what they wrote or may prefer to talk about their thoughts.

You may want to work together to create a poster entitled "Characteristics of a Well-Functioning Group." Such resources, which can be developed throughout the unit and the entire year, are useful for asking students to reflect on how their groups are working or to consider what role they can take to ensure that their group functions well. You might begin by asking students to privately list things they would see happening in a well-functioning group. Then have students volunteer ideas, while you record them on the board or chart paper, until their lists are depleted.

Key Questions

How did you know when you were done?

How confident were you in knowing you had solved the problem? Why were you so confident?

What mathematics was involved in this activity?

What else was mathematical about the ways your group worked?

Supplemental Activity

Counting Llama Houses (extension) asks students to identify the ways in which the houses in *Lonesome Llama* differed and then to determine how many different houses could have been created using these variations.

Role Reflections

Intent

This activity draws upon the *Lonesome Llama* experience in such a way that students recognize the various roles that members must assume in order for a group to function well.

Mathematics

IMP activities are designed to encourage a high degree of interaction, involving collaborative problem solving as well as engaging discussions of ideas, arguments, and presentations. For such a learner-centered environment to be effective, care must be taken to help students work in groups effectively. *Role Reflections* helps students become aware of these norms, or roles, for working in groups and why they are important. See "Assigning and Using Roles in Cooperative Groups" in the Overview to the Interactive Mathematics Program for more information.

Progression

Upon completion of *Lonesome Llama* (or as a filler while groups finish *Lonesome Llama*), students should privately reflect on the activity to identify actions taken by a student engaged in the various roles. A class discussion of responses helps to highlight the value of the variety of contributions that class members made to the successful completion of the activity.

Approximate Time

15 minutes

Classroom Organization

Individuals, then groups, followed by whole-class discussion

Doing the Activity

When the students in a group have determined that they are confident they know the *Lonesome Llama*, provide the following instructions. Ask each group member to think back to the activity and remember the various actions taken by their group members that helped them complete the task. Instruct students to organize their recollections using the roles listed in *Role Reflections*. This work should be done individually and privately.

Once *every* member of a group has completed identifying *one* person for each role, have the group members share what they recorded and what moment during the *Lonesome Llama* activity prompted them to identify each person.

Discussing and Debriefing the Activity

Once every group has spent at least some individual time identifying the moments that a member took on a particular role, bring the class together for a discussion. Ask for volunteers to share actions taken by fellow group members that moved them forward in their success to complete the *Lonesome Llama* activity. Ensure that both *task roles* and *socio-emotional roles* are shared.

Key Questions

Which role might someone in your group have performed that would have helped the group be successful with the activity?

Were their any roles that your group spent too much time attending to?

Communicating About Mathematics

Intent

The activities in *Communicating About Mathematics* build on the strands begun in *The Importance of Patterns,* while focusing on the written and oral communication of students' mathematical ideas.

Mathematics

One of the underlying principles of IMP is that doing mathematics involves offering ideas for analysis and critique as well as analyzing and critiquing the ideas offered by others. The process of convincing others of the validity of solutions and strategies, and of understanding and analyzing the solutions and strategies of others, is what mathematicians do when they present and publish proofs of theorems, solutions to problems, and new techniques for solving problems.

Progression

In *Communicating About Mathematics,* students will present their solutions to the first POW and begin work on the second one. They will explore a model for integer arithmetic and then write a primer on this model for others. This work will extend their understanding of arithmetic with integers and order of operations, building blocks of algebraic thinking. Students will also solve several problems that require careful reading of a mathematical situation and the creation and communication of solutions to an open-ended task. They will use a powerful problem-solving strategy called *working backward*.

Marcella's Bagels

1-2-3-4 Puzzle

Uncertain Answers

Extended Bagels

The Chefs' Hot and Cold Cubes

Do It the Chefs' Way

POW 2: Checkerboard Squares

You're the Chef

Marcella's Bagels

Intent

This activity engages students in a problem that requires close reading. Students also examine a POW-style write-up for this problem, which will support their write-ups for POW 1: *The Broken Eggs.*

Mathematics

Marcella's Bagels gives students an opportunity to use a variety of problem-solving strategies. They might guess at the original number of bagels, examine what happens when they work through the steps in the problem, and then revise their initial guesses accordingly.

The problem also lends itself to the powerful strategy of working backward. Thinking of the story as a movie, students can begin with the number of bagels Marcella has at the end and "run the movie backward," undoing each action she took and arriving at the number of bagels she had at the start. At each step, Marcella gives away half of her bagels plus 2, so in reverse she would add 2 and then double the total.

Progression

Students work backward or use other methods to analyze a complex word problem. They also examine a POW-style write-up for this problem.

Approximate Time

30 minutes

Classroom Organization

Groups

Materials

About 100 beans, counters, or similar items per group

Doing the Activity

You might introduce *Marcella's Bagels* by asking for volunteers to silently enact the events, with or without props, as you narrate the story. Afterward, ask students to restate the problem.

As you direct students to work in their groups on the activity, encourage them to use the materials available to help them think through the problem. If students express that they are beyond using objects like beans or counters, assure them that doing mathematics involves using whatever it takes—pencil and paper, calculators and computers, models and manipulatives—to understand a situation or an idea.

Discussing and Debriefing the Activity

Discuss the various methods students used to solve the problem.

How did you find the answer? How do you know your answer is correct?

The two most likely approaches will be (1) guessing the starting amount and then running the problem forward to see if the guess leads to the correct ending amount and (2) working backward.

Next, students will focus on how to communicate how they solved a problem like this and how to show and explain their solutions. Tell students that this is what they are asked to do in their write-ups for *POW 1: The Broken Eggs*.

Have students turn to the *POW-Style Write-up of Marcella's Bagels* at the end of the unit, which uses *Marcella's Bagels* to illustrate the POW write-up components. Ask them to read the example write-up on their own. When they have finished reading, ask, **What did you notice that is helpful in this write-up? What is missing? What isn't needed?**

Draw students' attention to the ways the writer used the components in the POW write-up. Ask, **How does the process used here differ from the solution?** In particular, bring out the observation that the write-up describes how the writer *thought about* the problem; it doesn't merely present the answer.

Key Questions

How did you find the answer?

How do you know your answer is correct?

What is helpful in this write-up?

What is missing?

What isn't needed?

How does the process used here differ from the solution?

Supplemental Activity

It's All Gone (reinforcement) is a variation on *Marcella's Bagels*, in which a man goes from store to store getting and spending money and winds up with no money in the end. Students are asked to determine how much money he had when he started.

1-2-3-4 Puzzle

Intent

1-2-3-4 Puzzle and its companion activity, *Uncertain Answers*, help students gain insight into the need for rules for order of operations and provide additional experience with the algebraic logic of graphing calculators.

Mathematics

Order of operations is a set of conventions that facilitate mathematical communication. By convention, arithmetic problems are worked out according to the following precedence rules:

- Simplify expressions within parentheses before combining them with expressions outside the parentheses.

- Within parentheses (or where no parentheses exist), do operations in this order: (1) Apply exponents to their bases. (2) Multiply and divide as the operations appear from left to right. (Neither operation has precedence over the other.) (3) Add and subtract as the operations appear from left to right. (Neither operation has precedence over the other.)

These precedence rules have been established to remove the ambiguity from the meaning of such written expressions as $3 \cdot 7 + 2^2$, which might otherwise be evaluated by multiplying 3 by 7, adding 2, and then squaring the result to obtain 529. Using the precedence rules above, the value of this expression is 25, because $2^2 = 4$, $3 \cdot 7 = 21$, and $21 + 4 = 25$.

Progression

This open-ended exploration highlights the importance of order-of-operations rules for communicating mathematically and gives students an opportunity to explore order of operations on the graphing calculator. The activity is also an ideal time to establish the conventional order of operations.

Approximate Time

30 minutes

Classroom Organization

Groups

Doing the Activity

As you begin planning for this activity, keep in mind these three components, which work well together: this activity, the companion activity *Uncertain Answers* (in which students examine order of operations on the graphing calculator), and a brief lecture on order of operations. One sequence is to have students begin the exploration of *1-2-3-4 Puzzle*, conduct the lecture, and assign *Uncertain Answers* for homework.

Many students enjoy this challenging puzzle. There are lots of ways to use 1, 2, 3, and 4 to generate each answer from 1 to 25. For example, $1 + 2 + 3 + 4 = 10$ and $3 + 2 \cdot 4 - 1 = 10$. It isn't necessary that all students find an expression for every number from 1 to 25; rather, just ask them to find as many 1-2-3-4 expressions as they can for 1 to 25. You might hang a poster in the room, with the numbers from 1 to 25 on it, and invite students to add new 1-2-3-4 expressions to it at any time.

Some key points to consider when orchestrating this activity:

- You might need to clarify the meaning of **factorial** and its position within the rules for order of operations. It has priority over the other operations, including exponentiation. For instance, $2 \cdot 3!$ means $2 \cdot (3!)$ not $(2 \cdot 3)!$. Using the $\wedge$ notation for exponents, $2\wedge3!$ is interpreted as $2\wedge(3!)$ not $(2\wedge3)!$.

- If any of the following graphing calculator basics were not discussed during *Calculator Exploration,* this activity may present opportunities to raise them. (It is not a requirement that all students know all these techniques at the conclusion of this activity.)

  Editing an expression that has been entered incorrectly (rather than starting over), including use of the INSERT key

  Entering an exponent using the $\wedge$ key

  Using parentheses and recognizing that braces (the { and } keys) and brackets (the [and] keys) do not work like parentheses

  Copying the last entry

  Using the answer to the last calculation as part of a new calculation

- Some students may have learned the acronym PEMDAS as a way to remember the order of operations. Unfortunately, this memory device reinforces the common misconception that multiplication is performed before division, and addition before subtraction. Reiterate that within each pair of operations, the operations are performed from left to right. For instance, in the expression $12 \div 6 \cdot 2$, the division is performed first.

- There is nothing wrong with inserting parentheses that aren't strictly required. People often do this in order to avoid any chance of confusion. For example, one might write the expression $5 \cdot 3 + 5 \cdot 7$ as $(5 \cdot 3) + (5 \cdot 7)$. Not only is the latter expression harder to misinterpret, it's also easier to see the intent at a glance.

This is an easy activity to engage students in. Begin by simply asking someone to volunteer a numeric expression using each of the numbers 1, 2, 3, and 4 and any operations they would like. Record their suggestion, and ask the class to calculate the result. Here, or whenever the possibility for multiple interpretations of an expression arises, is a good time to begin discussion of order of operations and the use of parentheses.

Ask for two or three more expressions, again instructing the class to calculate the results, and then wonder aloud, **Do you think we could create an expression for every number from 1 to 25?**

In their groups, students can productively explore for 15 or more minutes easily. At some stage—possibly after a break to introduce order of operations more formally and introduce the activity *Uncertain Answers*—gather the class to review the activity instructions. Reading the instructions will give students more ideas about the operations they can use. Many students won't have thought to use a square root, and few will be familiar with factorials.

Discussing and Debriefing the Activity

Students will be interested in the discussion of this activity in order to see expressions for numbers they haven't yet figured out. Have volunteers share expressions for answers that other students haven't found.

During the discussion, opportunities will arise to clarify order-of-operations rules. As they present themselves, ask the volunteer or the class to help rewrite the solution in the conventional form.

You might ask questions like the following to encourage volunteers to also talk about *how* they found their 1-2-3-4 expressions.

What methods did you use to find your expressions?

Did you proceed in numeric order or did you jump around?

Did you get an expression for one number by adjusting the expression for another?

Did you use any patterns that you saw in the expressions?

Order of Operations

Though many students have been exposed to **order-of-operations** rules, we treat the topic here as if some have not.

To introduce the topic, you might write arithmetic expressions involving several operations, such as these, on the board, and ask students to work on their own to find the value of each expression.

$$4 + 5 \cdot 3 + 1 \qquad\qquad 3 + 4^2$$
$$10 + 2 - 4 + 3 \qquad\qquad 2 + 4 \cdot 3^2$$
$$2 + 3(5 + 4) \qquad\qquad 12 \div 4 - 3$$

Some students may remember and apply the order-of-operations rules to get the correct answers, while others may never have learned the rules or may have forgotten them. Ask students to share their results for one or two of the expressions, and go through the details of their computations to demonstrate how the expressions can be interpreted in more than one way. Then point out that it would create great difficulties if more than one answer were correct. Mathematicians, scientists, and everyone who deals with numbers must communicate in writing, so there is a need for a set of rules that will govern how to interpret any apparent ambiguity in a problem.

Tell students that, by convention, arithmetic problems are worked out according to these rules:

- Simplify expressions within parentheses before combining them with expressions outside the parentheses.

- Within parentheses (or where no parentheses exist), do operations in this order:

  Apply exponents to their bases.

  Multiply and divide as the operations appear from left to right. (Neither operation has precedence over the other.)

  Add and subtract as the operations appear from left to right. (Neither operation has precedence over the other.)

Post the rules so that you and students can refer to them when needed. You might use a shortened version, such as the one that appears in *Uncertain Answers.*

Key Questions

Do you think we could create an expression for every number from 1 to 25?

What methods did you use to find these expressions?

Did you proceed in numeric order or did you jump around?

Did you get an expression for one number by adjusting the expression for another?

Did you use any patterns that you saw in the expressions?

Uncertain Answers

Intent

This activity gives students opportunities to gain insight into the need for rules for order of operations and helps to establish the conventional rules. This activity also gives them additional experience with the algebraic logic of graphing calculators.

Mathematics

This activity reinforces order of operations as students fix equations by inserting parentheses so that the resulting statements are correct.

Progression

This activity complements students' work in *1-2-3-4 Puzzle*.

Approximate Time

20 minutes for activity (at home or in class)

15 minutes for small-group discussion

Classroom Organization

Groups

Doing the Activity

This activity will require little or no introduction.

Discussing and Debriefing the Activity

You can give groups a few minutes to share their work on the assignment. Students should be able to resolve each other's difficulties within this group discussion.

If you see common errors as you circulate among groups, you may want to draw the class together for clarification, perhaps calling on individual students to explain a given idea. Clear up any conflicts by having students go through the problem one small step at a time.

Supplemental Activity

1-2-3-4 Varieties (reinforcement) adds a rule to those that students used in *1-2-3-4 Puzzle:* now the digits must appear in numeric order. In addition to finding expressions for the first 25 whole numbers, students are asked to find the greatest possible answer given these rules and to make up their own variations to the original activity.

Extended Bagels

Intent

In this activity, students explore what it means to extend a problem, which can help one gain a better understanding of the structure of a problem.

Mathematics

This activity extends *Marcella's Bagels* by posing the question of how altering the final number of bagels would change the initial number of bagels. This is a "functions" question: the starting number is a function of the ending number. Students are asked to find that functional relationship by trying several ending values, organizing their findings in an In-Out table, and then searching for a rule.

Progression

Students use backward reasoning or other methods to further investigate a complex word problem.

Approximate Time

15 minutes

Classroom Organization

Groups, followed by whole-class discussion

Materials

About 100 beans, counters, or similar items per group

Doing the Activity

Introduce the question that frames this extension to the original *Marcella's Bagels* problem: **How does the solution to *Marcella's Bagels* depend on the number of bagels Marcella has when she gets home?**

Monitor group interaction. Encourage students to share ideas and to make sure everyone has an opportunity to contribute his or her ideas.

Discussing and Debriefing the Activity

Have a group or student volunteers share the ideas they pursued and what they learned about the problem. An In-Out table with all their data might look like this.

Number of bagels when Marcella gets home	Number of bagels Marcella started with
0	28
1	36
2	44
3	52
4	60
5	68

It is not crucial that students develop a rule to describe this relationship. If they haven't found a rule, you might post the In-Out table and invite students to continue to think about a rule and bring their ideas to you when they have time.

Conclude this activity with some discussion of the idea that each Problem of the Week (POW) requires students to write out, and sometimes explore, an extension to the original problem. Mention that mathematics is at least as much about creating interesting questions as it is about answering them.

Key Question

How does the solution to *Marcella's Bagels* depend on the number of bagels Marcella has when she gets home?

The Chefs' Hot and Cold Cubes

Intent

The "hot and cold cubes" sequence of activities offer a model for the operations of integer arithmetic, key tools in high school mathematics. This activity reviews some basics about negative numbers and reaffirms some conventions before students begin to make sense of the "hot and cold cubes" model for integer arithmetic.

Mathematics

The IMP program assumes that most students have had prior exposure to negative numbers and have been taught—but may not remember or understand—basic rules for arithmetic with integers. In this set of activities, students are introduced to a model that embodies these rules and serves as a metaphor for thinking about integer arithmetic. Rather than simply reviewing the rules for such arithmetic, this model provides a frame of reference for the rules and will allow students, if necessary, to reconstruct the rules for themselves in the future.

The basic operations for **natural numbers** are defined, at least intuitively, in terms of putting sets of objects together and taking objects away from a set, but this definition doesn't make sense for negative numbers. In moving from **whole numbers** to **integers**, numbers are no longer simply a magnitude, but also a direction. Treating an integer as merely an opposite of a whole number, as do such commonly memorized rules as "subtracting a negative is the same as adding a positive," does not encourage a more powerful understanding of integer. The hot and cold cubes model emphasizes both the magnitude and the direction of an integer and encourages awareness of the meaning of the operation involved.

Progression

The activity begins with discussion of the need to justify solutions when doing integer arithmetic and is followed by a brief review of notation, language, and conventions. Students are then introduced to the model and, in their groups, perform some integer arithmetic to help make sense of the model. They are asked to translate the chefs' moves (using hot and cold cubes to change the temperature in the cauldron) into integer arithmetic and to translate integer arithmetic into chefs' moves.

Approximate Time

35 minutes

Classroom Organization

Groups

Materials

Manipulatives, such as cubes or tiles, in two colors

Doing the Activity

Many students have probably been exposed to the basics of computing with negative numbers and may be reluctant to learn another way of thinking about the process. The hot and cold cubes model, however, will help them to understand *why* the rules work. It offers a frame of reference for the rules and will allow students to reconstruct the rules for themselves in the future.

Two quick questions can provide information on students' prior knowledge.

What is the answer to $(^-3)(^-5)$? How do you know your answer is right?

What is the answer to $^-3 + ^-5$? How do you know your answer is right?

You will likely receive a variety of solutions; record them all on the board. Remind students that they should be able to state why their answers are correct. Some students may be able to apply the rules to find the correct answers, but many will have trouble explaining why the product of two negative numbers is positive while the sum of two negative numbers is negative. The intent of this introductory challenge is to convince students that they still have something to learn about working with negative numbers and that it might be worthwhile to have an approach that doesn't rely on memorizing rules.

Review the notation and terminology of positive and negative numbers. These activities use the "raised sign" notation, such as $^+5$ and $^-7$. These should be read as "positive five" and "negative seven" not "plus five" or "minus seven." Using clearly defined terminology helps to distinguish between positive and negative *numbers* and the *operations* of addition and subtraction.

Once these activities are completed, the student book reverts to the standard notation conventions: positive numbers typically include no sign, and negative numbers are denoted with the same symbol as subtraction. Tell students that this convention is typical in most contexts.

Also review the following terminology and notation with students.

- The *sign* of a number indicates whether it is positive or negative. Zero is considered neither positive nor negative.

- The numbers in a pair such as $^+3$ and $^-3$ are sometimes called *opposites*. That is, $^-3$ is the opposite of $^+3$, and $^+3$ is the opposite of $^-3$.

- The word **integer** refers to a number that is zero, a natural number, or the opposite of a natural number. (The **natural numbers** are positive whole numbers: 1, 2, 3, 4 and so on.) Thus the set of integers is

$$\{\ldots, ^-3, ^-2, ^-1, 0, ^+1, ^+2, ^+3, \ldots\}$$

- The *number line* is a way to picture both positive and negative numbers. By convention, positive numbers are on the right and negative numbers are on the left; numbers are considered to get larger as one moves to the right on the number line. Thus, for example, $^+5 > ^-8$ and $^-7 < ^-3$.

```
   ⁻4   ⁻3   ⁻2   ⁻1    0    ⁺1   ⁺2   ⁺3   ⁺4
←——+————+————+————+————+————+————+————+————+——→
```

Have students read the introduction to *The Chefs' Hot and Cold Cubes* and the first five paragraphs of The Story.

Introduce students to the manipulatives—such as two colors of cubes or tiles—for representing hot and cold cubes. Ask groups to use their manipulatives to create several cauldrons, each representing a temperature of 0°, to introduce the idea that a hot cube and a cold cube "cancel out" one another.

Have groups read the next paragraph (beginning "For each hot cube . . .") and then create cauldrons for other specific temperatures, such as ⁺5° or ⁻3°. The idea is for students to get a sense of the cancellation mechanism and to see that a given temperature can be represented in many ways.

After this introduction, let students read the rest of The Story individually and then work in their groups on the questions.

As groups work, you may want to emphasize that the equations and arithmetic expressions focus on the *change* in temperature and not on the temperature itself.

Some students may be confused by the fact that the same notation is used in different ways. For instance, ⁺5 can mean "add five bunches of a certain number of hot or cold cubes" (as in ⁺5 • ⁺20 = ⁺100) or "a bunch containing five hot cubes." If this comes up, acknowledge that this part of the model is something they may have to pay extra attention to. It is similar to the dual meaning in multiplication of whole numbers, in which 5 • 3 can mean "5 groups with 3 objects in each group" or "3 groups with 5 objects in each group," with 5 representing either the number of groups or the size of each group.

Discussing and Debriefing the Activity

This activity will likely not require a formal debriefing.

Key Questions

What is the answer to (⁻3)(⁻5)? How do you know your answer is right?

How might you represent the situation with objects?

Supplemental Activity

Positive and Negative Ideas (extension) extends the work with hot and cold cubes and asks students to consider other ways they might model integer arithmetic.

Do It the Chefs' Way

Intent

This activity gives students more experience with the model of hot and cold cubes for integer arithmetic.

Mathematics

Students use the hot and cold cubes model to understand arithmetic with integers. This is also a good time to introduce the concept of **absolute value** and to explore patterns in operations with integers.

Progression

Students have worked in their groups to make sense of the "hot and cold cubes" model. Now they will spend some individual time practicing with and confirming their understanding of the model. After comparing their work with one another and discussing questions that arise, students review a few more basic ideas related to operations with integers.

Approximate Time

20 minutes for activity

30 minutes for discussion

Classroom Organization

Individuals, followed by whole-class discussion

Materials

Manipulatives, such as cubes or tiles, in two colors

Doing the Activity

To introduce the activity, emphasize that students are to explain each expression in terms of the "hot and cold cubes" model.

Discussing and Debriefing the Activity

Have a volunteer from each group explain one of the answers. Insist on the use of the "hot and cold cubes" model, even if the student prefers to quote arithmetic rules.

For example, for Question 4, a student should say something like, "The term ⁻4 (read "negative four") means the chefs put in four cold cubes, which lowered the temperature 4°. The –⁺6 (read "subtract positive six") means they took out six hot cubes, which lowered the temperature 6°. Altogether the temperature went down 10°. In terms of the model, you would write ⁻4 – ⁺6 = ⁻10."

Some students may comment that taking out hot cubes has the same effect as putting in cold cubes. Such alternative explanations of the expressions, in terms of the model, should be encouraged.

In a problem like $^+5 \bullet {}^-2$ (Question 3), students may say that the answer is negative "because a positive times a negative makes a negative." Encourage a reinterpretation of such an expression in terms of the model. For example, they might say, "This is as if the chefs were putting in bunches of cold cubes, which lowers the temperature."

If some students continue to resist learning the model, insisting that they can get the answers more easily from rules, mention that part of learning mathematics is being able to explain it and that being able to explain simple situations like this is good practice for explaining more complex problems later. Point out that students do not have to use the model for every computation they do now or in the future, but they should be prepared to justify their work in terms of the model when asked to do so.

Absolute Value

Working with the hot and cold cubes model is an ideal opportunity to introduce the term **absolute value**. Tell students that the absolute value of an integer is the number of cubes it represents. Help them to see that any integer except zero is a combination of a sign and an absolute value.

Also introduce the notation for absolute value through examples, such as $|5| = 5$, $|^-7| = 7$, and $|^-0| = 0$.

Ask students, What's the difference between the operation of subtraction and the negative sign?

If students develop their own general rules about the relationship between sign and operation, such as "adding a negative gives the same result as subtracting a positive," that's fine. However, tell them that familiarity with the hot and cold cubes model will give them something to fall back on if they happen to forget the rule.

At this point you can announce that you (and their books) will generally omit the raised plus sign prefix for positive numbers and will write the negative sign the same way as a subtraction sign.

Explain that in order to avoid seeming to write two arithmetic operation symbols next to each other, it is common to insert parentheses. For example, instead of $10 + -7$ or $8 - -4$, we might write $10 + (-7)$ or $8 - (-4)$. Also, for $5 \bullet -3$, we might write $5 \bullet (-3)$, but we often omit the multiplication sign when there are parentheses to indicate multiplication: $5(-3)$. Students are probably familiar with the use of parentheses for multiplication in such expressions as $5(2 + 7)$, but may not have seen it used in conjunction with a symbol immediately inside the parentheses that could be interpreted as an arithmetic operation, such as $5(-2 + 7)$.

Patterns in Integer Arithmetic

At this point, you may want to present the following pattern approach to operations with integers. Begin by writing this sequence of addition equations.

$7 + 3 = 10$

$7 + 2 = 9$

$7 + 1 = 8$

$7 + 0 = 7$

$7 + (-1) = ?$

Ask students to look for a pattern and use it to explain what number belongs in place of the question mark. **What should come next in this sequence?** Presumably they will see that the sequence of answers suggests that $7 + (-1)$ should equal 6.

Continue with $7 + (-2) = ?$ and similar problems. Ask students what is happening. They should see that as negative numbers of greater magnitude are added, the resulting sum gets smaller.

You may want to continue through $7 + (-7)$ and on into examples that give a negative sum, such as $7 + (-8)$.

After the pattern has been described, ask the class whether this pattern gives the same answers as the hot and cold cubes model. Students should be able to explain how to get the same results from the model.

Present the next series of equations, which relate to subtraction, and ask, **What should come next in this sequence?**

$7 - 5 = 2$

$7 - 6 = 1$

$7 - 7 = 0$

$7 - 8 = ?$

From the continuation of this pattern, students should see that if a greater number is subtracted from a lesser number, the result is a negative number. Some may also notice that the result is the opposite of the result when the two numbers are reversed.

Continue the pattern above in the opposite direction, subtracting smaller and smaller numbers from 7. **What should come next in this sequence?**

$$7 - 5 = 2$$
$$7 - 4 = 3$$
$$7 - 3 = 4$$
$$7 - 2 = 5$$
$$7 - 1 = 6$$
$$7 - 0 = 7$$
$$7 - (-1) = ?$$

Students should notice that as the number being subtracted grows smaller, the result increases. Some may recognize that subtracting a negative number gives the same result as adding the corresponding positive number, so that there is a related addition equation for each subtraction equation. For example, the subtraction equation $7 - (-5) = 12$ relates to the addition equation $7 + 5 = 12$.

Finally, have students look for patterns in the products of integers, as in this sequence.

$$6 \cdot 3 = 18$$
$$6 \cdot 2 = 12$$
$$6 \cdot 1 = 6$$
$$6 \cdot 0 = 0$$
$$6 \cdot (-1) = ?$$

Students should observe that as the second factor decreases by 1, the products decrease by 6.

Key Questions

What's the difference between the operation of subtraction and the negative sign?

What should come next in this sequence?

How does this pattern relate to the hot and cold cubes model?

POW 2: Checkerboard Squares

Intent

This is the second POW of the course. The primary purpose of this, as for all POWs, is to give students the opportunity to solve a significant problem outside of class, to generalize their solutions, and to prepare a formal written account of their work.

Mathematics

This POW will draw on and strengthen students' ability to visualize geometrically, to collect and organize a complex set of information, and to generalize their solutions. Students are asked to generalize their methods for counting the number of squares of different sizes on an 8-by-8 checkerboard to produce a method for counting the squares on a board with dimensions *n* by *n*.

This activity is connected to an important mathematical theme of this unit: patterns. In counting all the squares of different sizes on a checkerboard, students will have to be systematic to ensure they have accounted for all the squares. To do so, they will have to recognize patterns in the locations of squares of various sizes.

Progression

This POW is posed toward the end of *Communicating About Mathematics*, and students will work on it into *Investigations*. Unlike *The Broken Eggs*, the student book includes no follow-up activities in support of the various components of students' write-ups of this POW. Students will be introduced to summation notation in another activity, *Add It Up*, as they are working outside of class on this POW. This notation can then be brought into *Checkerboard Squares* as appropriate.

Approximate Time

10 minutes for introduction

10 minutes for discussion

1 to 3 hours for activity (at home)

20 minutes for presentations

Classroom Organization

Whole-class introduction, concluding with presentations and class discussion

Doing the Activity

The first part of every POW write-up is the student's statement of the problem. In addition to helping students learn how to write a problem statement, having students work on and share their problem statements soon after the POW is assigned will help clarify for many students what the problem is.

Announce when the write-up is due. Solicit presenters immediately, or nearer the due date, reminding the presenters of basic expectations and providing them with transparencies and pens.

Discussing and Debriefing the Activity

Discussion of this POW can possibly extend over two days.

Before students turn in their write-ups, you might offer them an opportunity to review the work of other students. This is their second POW, so they will have formed some idea of what is expected, but seeing each other's work may be of great value.

As they read other students' work, you might have students focus on what makes a good paper, what makes an adequate paper, and what makes a poor paper. After the sharing of POW write-ups is complete, you might want to ask students to do focused free-writing on this topic: **What makes a good POW write-up?** (see "Focused Free-Writing" in the Overview to the Interactive Mathematics Program). After they have written for about five minutes, let students share their ideas. They can read aloud from their written work or simply discuss what they wrote about.

Have the assigned students give their presentations, limiting each to about five minutes. Encourage presenters to speak about their investigation process at least as much as they speak about their findings. When findings overlap, presenters may wish to emphasize slight nuances they saw, questions they explored, and the like.

In the discussion that grows out of the presentations, focus on the patterns that students have discovered. Bring out that finding patterns helps us to analyze mathematical situations.

Student interest may offer opportunities to extend the exploration. For example, this POW lends itself to trying to explain *why* the square numbers appear. Students, or you, may raise such questions as these: **Why is the number of squares of each size itself a square number? Why is it the particular square that it is?**

The problem can also be another opportunity to use summation notation. You might inquire, **Do you see a way to express your findings using summation notation?**

Key Questions

Why is the number of squares of each size itself a square number?

Why is it the particular square that it is?

Do you see a way to express your findings using summation notation?

Supplemental Activities

Different Kinds of Checkerboards (extension) is a follow-up to *POW 2: Checkerboard Squares* in which students find the number of squares on nonsquare checkerboards and search for a general rule for checkerboards of dimensions *m* by *n*.

Lots of Squares (extension) is a substantial investigation in which students are asked to divide a square into different numbers of smaller squares. The goal is to determine which numbers of smaller squares are impossible and which are possible and to prove their results.

You're the Chef

Intent

This activity concludes work with the "hot and cold cubes" model for integer arithmetic, for now. This activity requires students not only to make meaning of the model but to create a thorough explanation of the model, using examples. Their write-ups of this activity will be part of their *Patterns* portfolios.

Mathematics

Throughout the rest of the program, students are expected to use negative numbers where appropriate.

Progression

Students will use their previous experiences to put the hot and cold cubes model into their own words, including selecting several examples to provide a complete explanation of the model.

Approximate Time

5 minutes for introduction

30 minutes for activity (at home or in class)

10 minutes for discussion

Classroom Organization

Individuals, followed by whole-class discussion

Doing the Activity

Have students read the activity. Advise them to be thoughtful about including sufficient examples in their manuals so that the new assistant chef will know how to read, interpret, and write recipes.

Tell students that they will include their manuals in their portfolios, which they will create at the end of this unit.

Discussing and Debriefing the Activity

You might have a few students read their papers to the class. Encourage students to identify parts of the manual they struggled to write. Ask for other students to comment, sharing how they figured out how to write that section or even reading the passage aloud.

Rather than collecting the assignment now, you may want to discuss the ideas and then let students turn in revised versions tomorrow.

Supplemental Activity

Chef Divisions asks students to model division with hot and cold cubes.

Investigations

Intent

Investigations contains several rich mathematical explorations that offer students opportunities to employ the tools and techniques they developed in *The Importance of Patterns* and *Communicating About Mathematics*.

Mathematics

The two groups of investigations in *Investigations* draw on important ideas from number theory and plane geometry. In the first set, students will propose, test, and justify conjectures (and find counterexamples) about sums of whole numbers. They will be introduced to some formal notation for expressing conjectures and will continue to explore the notions of existence and uniqueness of solutions. In the second set of activities, students will apply their developing skill at using In-Out tables to find patterns in the relationship between the sides and angle measures of polygons.

Progression

Investigation of Consecutive Sums

> Consecutive Sums
>
> Add It Up
>
> Group Reflection
>
> That's Odd!

Investigation of Polygon Angles

> Pattern Block Investigations
>
> Degree Discovery
>
> Polygon Angles
>
> An Angular Summary

Consecutive Sums

Intent

The core for a series of activities that open *Investigations, Consecutive Sums* poses an open-ended situation in which students are encouraged to make and test conjectures, construct proofs, and find counterexamples. This activity helps to establish a classroom environment of student-student interaction during the exploration of a challenging mathematical investigation.

Mathematics

Students examine complex, open-ended mathematical questions, develop and test ideas, write proofs using logical reasoning and algebraic notation, and disprove **conjectures** using **counterexamples**.

Students are also introduced to summation notation. Consecutive sums are defined as sums of consecutive natural numbers, such as $6 + 7 + 8 + 9 = 30$, $35 + 36 = 71$, and $1 + 2 + 4 + \cdots + 10 = 55$. The third example can be written, using summation notation, as $\sum_{i=1}^{10} i = 55$.

Progression

This activity requires a brief introduction, followed by a significant amount of time working in small groups to investigate, prepare posters, and present findings to the class. Split the activity over at least two days, assigning the individual activity *Add It Up* in the break. In *Group Reflection*, students reflect on the nature of the activity and the ways in which they worked together during the investigation.

Approximate Time

85 minutes

Classroom Organization

Groups

Doing the Activity

Students' task is to explore patterns in consecutive sums and to create a poster summarizing their work, including descriptions of confirmed, disproved, and still-open conjectures.

Once you ensure that students understand what consecutive sums are and what they are being asked to do, the best way for students to begin is to "try stuff." For the purpose of this activity, a **conjecture** might be defined as a "guess based on some evidence."

To introduce the activity, you might spark student interest by offering a few examples of consecutive sums and posing a challenging question, such as one of those given below.

This activity will be explored over at least two days. The end of the first day's work is a good time to refocus groups on the products they are to create: posters that summarize their results and contain summary statements of the patterns they have observed. Remind students that they will classify patterns as conjectures, certainties (statements that are always true), and false conjectures.

Encourage groups to record clearly worded summary statements about what they think the pattern is. Offer an example of a clear summary statement, such as "Every number can be written as a consecutive sum." Tell students that while this statement may or may not be true, it is the type of statement you are looking for.

The following are some of the questions that groups might investigate.

What numbers can be written as consecutive sums?

What numbers can be written as more than one consecutive sum?

Are there patterns to the answers to consecutive sums that are two terms long (such as 4 + 5), three terms long, or four terms long?

If groups have gathered some information but are not seeing any patterns, suggest that they try to reorganize the information in a way that might make patterns more visible.

As the exploration draws to a close, circulate to help groups focus on their summary statements. Following are some possible summary statements.

- Every odd number greater than 1 can be written as a consecutive sum of two terms. (This particular statement is the subject of the activity *That's Odd!*) Because only positive whole numbers are permitted in the activity, 1 itself cannot be written as a consecutive sum.

- The numbers 1, 2, 4, 8, 16, . . . (powers of 2) cannot be written as consecutive sums.

- The numbers that cannot be written as consecutive sums are all even. (This statement is incorrect, because 1 is odd but cannot be written as a consecutive sum. It can be written as 0 + 1, but the activity allows only positive terms, not 0.)

- Every third number—that is, every multiple of 3—except 3 itself can be written as a consecutive sum of three terms. (The number 3 is 0 + 1 + 2, but again, 0 is not permitted.)

Discussing and Debriefing the Activity

Once groups have displayed their posters, review and discuss this collection of conjectures and summary statements. Ask a member of each group to state one of the patterns that the group found that hasn't yet been mentioned. Continue until no group has summary statements that haven't already been mentioned.

It may work best to have all the statements read before getting into discussion of or challenges to any of them. When ready, invite students to comment on the summary statements of other groups. They may have facts that contradict a given statement, or they may simply question whether a given generalization is valid.

Introduce the word **counterexample** in the context of these summary statements by asking whether there are any cases in which a generalization doesn't hold. (If no one offers one, suggest one yourself.) For example, the summary statement "If a number can be written in three or more ways as a consecutive sum, then it must be odd" is false, and 30 is a counterexample. Although 30 fits the condition that "it can be written in three or more ways as a consecutive sum," it doesn't have the property "it must be odd."

On the basis of this discussion, the class may eliminate or confirm some of the summary statements, while others will remain conjectures. For example, the statement "Powers of 2 cannot be written as consecutive sums of positive whole numbers," though a true statement, will probably remain unproven at this time.

Key Questions

What numbers can be written as consecutive sums?

What numbers can be written as more than one consecutive sum?

If a number can be written as a consecutive sum, is that consecutive sum unique?

What numbers are not answers to some consecutive sum? Are there patterns in these numbers?

Are there patterns to the answers to consecutive sums that are two terms long (such as 4 + 5), three terms long, or four terms long?

Supplemental Activities

Three in a Row (extension) offers students an opportunity to explore sums of three consecutive numbers as well as sums of other lengths.

Any Old Sum (extension) asks students to examine sums that are not consecutive.

Add It Up

Intent

This activity introduces the mathematical symbol for summation notation. Students begin to understand the utility of this notation by working with both numeric and geometric examples.

Mathematics

One of the challenges of secondary mathematics teaching is helping students to understand the notational systems used to express complex ideas in a compact form. This activity introduces one such system, summation notation, and offers students opportunities to start to make sense of it.

Progression

This activity serves as a useful way to break up student work on *Consecutive Sums*. Students may elect to utilize summation notation in their posters for *Consecutive Sums* and in *POW 2: Checkerboard Squares*.

Approximate Time

10 minutes for introduction

20 minutes for activity (at home or in class)

10 minutes for discussion

Classroom Organization

Individuals, followed by whole-class discussion

Doing the Activity

Introduce the activity with a multiterm example of a consecutive sum, such as $3 + 4 + 5 + 6 + 7 + 8 + 9$. Demonstrate that there is a shorthand way for writing such sums: $\sum_{i=3}^{9} i$. Explain that this symbol is an uppercase letter in the Greek alphabet, called *sigma*, and that the expression is read, "The summation, from *i* equals 3 to 9, of *i*." Invite students to articulate the connection between the shorthand and the full expression.

Explain that the letter *i* is called a *dummy variable* and that any letter would work. The expression $\sum_{t=3}^{9} t$ means exactly the same thing as $\sum_{i=3}^{9} i$.

Use a more complex example to illustrate in detail how this notation works. For example, ask students what they think this expression means.

$$\sum_{w=3}^{7} \left(w^2 + 2 \right)$$

How can you "act out" the process described by this summation expression? Help students act out the process.

First, w is 3, so the first term is $3^2 + 2$.

Then, w is 4, so the next term is $4^2 + 2$.

Then, w is 5, so the next term is $5^2 + 2$.

Then, w is 6, so the next term is $6^2 + 2$.

Finally, w is 7, so the next term is $7^2 + 2$.

Since the symbol Σ indicates summation, these terms must be added together. In other words, the notation represents the expression.

$$(3^2 + 2) + (4^2 + 2) + (5^2 + 2) + (6^2 + 2) + (7^2 + 2)$$

Point out that although this example does not give a consecutive sum, the values for w are a sequence of consecutive numbers.

The mechanics of summation notation are summarized in the student activity. Students will work with this notation in geometric as well as in purely numeric contexts. Don't get bogged down on mastery of the notation, it is intended only as a tool to help students express their ideas.

You may want to introduce the use of ellipsis notation, such as writing $1 + 2 + \cdots + 100$ for the sum of the whole numbers from 1 to 100

Discussing and Debriefing the Activity

Give students an opportunity to share responses and ask questions of one another.

For Question 2, students will likely see the picture in terms of the sum $1 + 2 + 3 + 4$ and produce an expression like $\sum_{i=1}^{4} i$.

The expressions for Question 3 can be written in various ways. Question 3c is especially likely to lead to different answers, such as $\sum_{t=2}^{6}(3t+2)$ and $\sum_{j=3}^{7}(3j-1)$. You can leave this question open if students cannot find a way to write the expression using summation notation.

For Question 4, the diagram suggests the idea of a sum of squares and can be expressed as $\sum_{n=1}^{5} n^2$.

Students' facility with summation notation will increase as they find situations where it is useful.

Key Question

How can you "act out" the process described by this summation expression?

Group Reflection

Intent

This activity is an opportunity for students to thoughtfully consider how they participate in groups.

Mathematics

In an open-ended exploration such as *Consecutive Sums,* productive group work is quite helpful. In this activity, students reflect on their experiences as members of a collaborative problem-solving group.

Progression

Students reflect on their participation in the open-ended activity *Consecutive Sums* to identify norms for productive interaction, to assess their level of participation, and to consider ways that they can draw others into the group.

Approximate Time

10 minutes for introduction

20 minutes for activity (at home or in class)

15 minutes for discussion

Classroom Organization

Whole class, then individuals, followed by whole-class discussion

Doing the Activity

Once students have completed *Consecutive Sums,* you might introduce this activity by asking them to think about this question: **What makes a group work well?**

Have students do focused free-writing on this topic. (See "Focused Free-Writing" in the Overview to the Interactive Mathematics Program.) Briefly explain the expectations of focused free-writing, if you haven't already. There are several key points to make.

- Their focused free-writing will not be collected (though students may have the opportunity to read some of it aloud or just share their ideas).

- Students should write, write, write—a sort of stream of consciousness.

- Students should try to stick with the assigned focus.

- Explain that it is better to write things like "I can't think of what to say" than to stop writing completely.

After a few minutes, have volunteers share their ideas, either reading what they wrote or just talking about their thoughts. If students are reluctant to share their ideas, you might prompt some conversation by asking questions such as the following.

Describe a time when you thought your group was working especially well together, when you were all achieving more than you could alone.

Describe a time when your group might have been able to be more productive. What was going on?

What strategies have you noticed you or other group members using that brought the group back together and that helped you be more successful?

Tell students that group work often does not come naturally and may not be a great experience if group members aren't thoughtful about how they are working and interacting. Learning to work well in groups is not necessarily an easy thing. However, as long as students get involved in their groups and respect the other members, they should see improvement.

Tell students that in this activity, they will think more specifically about the actions people take in groups.

Discussing and Debriefing the Activity

Let students share what they want about this activity. Such open discussion is important for developing a classroom climate in which students are willing to share their ideas and opinions.

To break the ice, you may want to talk about your own experiences working in groups. This is also an appropriate time to return to (and possibly edit) the posted classroom norms, group roles, or the Characteristics of a Well-Functioning Group poster from *Lonesome Llama*.

You may wish to collect and read these assignments, not for grading, but to gain insight into the dynamics of your classroom and the attitudes of your students.

Key Questions

Describe a time when you thought your group was working especially well together, when you were all achieving more than you could alone.

Describe a time when your group might have been able to be more productive. What was going on?

What strategies have you noticed you or other group members using that brought the group back together and that helped you be more successful?

Supplemental Activity

Getting Involved (reinforcement) asks students to reflect on a situation in which one person in a group is not contributing.

That's Odd!

Intent

The purpose of this activity is not for students to learn a proof that odd numbers greater than 1 can be written as a sum of two consecutive numbers. Rather, it is intended to help students do the following:

- Begin to learn what a proof is
- Learn to distinguish between specific examples and a general argument
- Gain experience in communicating complex, abstract ideas
- Become familiar with a more precise way of thinking than they may have encountered before

Mathematics

This activity challenges students to evaluate whether a conjecture is true or false. The conjecture in question is "If an odd number is greater than 1, then it can be written as the sum of two consecutive numbers." Students are asked to find a **counterexample** if they think the statement is false or to devise a set of instructions for writing any odd number greater than 1 as a sum of two consecutive numbers if they think the statement is true.

Progression

This activity should follow the conclusion of *Consecutive Sums*. Students work alone to consider a conjecture likely to have been made during that activity. Follow-up discussion will introduce the concept of a **proof**.

Approximate Time

15 minutes for activity (at home or in class)

10 minutes for discussion

Classroom Organization

Individuals, followed by whole-class discussion

Doing the Activity

Introduce the activity by asking, Is the conjecture in the activity true? How confident are you about your answer?

Clarify the instructions for students so that they understand what is expected of them.

Discussing and Debriefing the Activity

Begin the discussion by asking the class again whether they think the conjecture is true. Then ask how confident they are that it is true for *every* odd number greater than 1. Most students will likely be fairly sure that it is always true, but encourage skeptics to voice their opinions.

Ask for volunteers to share any instructions they developed for writing an odd number as a sum of two consecutive numbers, and have them illustrate their methods using specific examples. If the class is at a loss about how to do this, you might ask a series of questions, such as, **How would you write 397 as the sum of two consecutive numbers? How would you write 4913 as the sum of two consecutive numbers? How would you write 157,681 as the sum of two consecutive numbers?**

Encourage students to explain how to find the pair of consecutive integers in each case, as this is key to developing a general argument.

There are several ways to describe the general process; elicit as many as possible from your students. Here are some commonly suggested procedures.

- Subtract 1 from the odd number to get an even number. Divide this even number by 2. That quotient and the next number are the desired consecutive numbers.

- Divide the odd number by 2, getting "something and a half." The whole numbers just above and below this mixed number are the desired consecutive numbers.

- Add 1 to the odd number to get an even number. Divide this even number by 2. That quotient and the previous number are the desired consecutive numbers.

Careful examination of any of these methods will show that they don't work if the initial odd number is 1, because one of the numbers in the consecutive sum will be 0 rather than a positive whole number as required.

Whichever methods students suggest, ask them to explain how they know that a given method works. For example, for the first procedure above, you might ask how students know that subtracting 1 from an odd number gives an even number. The best response to this question would refer to a definition of the term *odd*. That is, students should recognize that, ultimately, they can't say anything for sure about odd numbers unless they begin with a clear definition. Similarly, ask how students know that dividing an even number by 2 gives a whole-number result. Again, encourage them to see that the answer to this challenge depends on having a precise definition of the term *even*. It is not necessary to go into formalities about the meaning of the terms *odd* and *even*. What is important is recognizing the value of having a precise definition if one is to give a complete proof.

Use the discussion to help bring out the difference between a collection of examples of a phenomenon and a legitimate general proof. A proof does not need to use algebraic symbols. For example, when appropriate and precise definitions are given for *odd* and *even,* the arguments above constitute completely legitimate proofs that every odd number can be written as a consecutive sum with two terms. Help students to see that these arguments are better than only giving a few examples such as 23 = 11 + 12 and 47 = 23 + 24.

Each procedure listed above demonstrates that every odd number is expressible as a consecutive sum of two terms by showing *how to do it,* that is, how to find the two terms. Such how-to arguments are considered legitimate proofs and are known as *constructive proofs.*

Algebraic symbols do sometimes help students understand a situation, and your students may be able to express their arguments symbolically. For example, if you suggest using *n* for the number obtained after subtracting 1 and dividing by 2, students can probably write the next number as *n* + 1.

You might extend this problem by encouraging students to express each method using a general equation. For instance, the first method listed above can be expressed by the equation $N = \dfrac{N-1}{2} + \left(\dfrac{N-1}{2} + 1 \right)$. Students can be asked to explain why, if *N* is odd and greater than 1, both terms must be positive integers.

Key Questions

Is the conjecture in the activity true? How confident are you about your answer?

How would you write 397 as the sum of two consecutive numbers?

How would you write 4913 as the sum of two consecutive numbers?

How would you write 157,681 as the sum of two consecutive numbers?

Supplemental Activities

The General Theory of Consecutive Sums (extension), the final activity in this group, asks students to explore consecutive sums of integers.

Infinite Proof (extension) asks students to prove that the square of every odd number is odd and that every prime number greater than 10 must have 1, 3, 7, or 9 as its units digit.

Pattern Block Investigations

Intent

This activity introduces students to pattern blocks, determining angle measures, and learning to use a protractor. Some students will already know how to measure angles with a protractor, and some will benefit from a refresher.

Mathematics

Pattern blocks are polygons that share side and angle relationships. By fitting and stacking these blocks, students can observe many geometric relationships inherent in these special manipulatives.

The equilateral triangle, the square, both parallelograms, and the hexagon have the same side lengths. The trapezoid has three sides of that length and one side twice that length.

Two triangles cover the blue parallelogram, three triangles cover the trapezoid, and six triangles cover the hexagon.

Three blue parallelograms, or two trapezoids, cover the hexagon.

The large angles in the blue parallelogram are twice the size of that figure's small angles.

The small angles in the tan parallelogram are half the size of the small angles in the blue parallelogram.

Students will also deduce the sizes of the interior angles of these polygons by examining the relationships among the polygons. Finally, they will use these known angle measures to figure out how to measure angles using a protractor.

The division of a complete turn into 360 equal parts is quite ancient and is often attributed to the Babylonians, whose number system was based on 60 and for whom the number 360 played an important role.

Progression

During the next sequence of activities, students will learn about the concept of angle. *Pattern Block Investigations* introduces students to pattern blocks, a manipulative tool that they will use in the development of this concept. In Part I of this activity, students create pattern block designs and focus their attention on the point formed by the blocks' vertices. In Part II, based on the idea that a full turn is 360 degrees, students deduce the sizes of all angles of all the blocks. Finally, in Part III they trace the blocks, extend their sides, and then use their deductions of angle sizes to learn to measure angles using a protractor.

Approximate Time

40 minutes

Classroom Organization

Groups

Materials

Pattern blocks

Overhead pattern blocks (optional)

Doing the Activity

When students encounter a new manipulative, they often need time to explore its properties and possibilities. Begin by providing groups with a large set of pattern blocks and encouraging a few minutes of exploration. As students explore, review the names for the various blocks: *triangle, hexagon, parallelogram* (or *diamond*), *square*, and *trapezoid*. You can refer to the two different parallelograms by shape (*wide* and *thin*) or by color (*blue* and *tan*). Also introduce the general term **polygon** as well as the term **quadrilateral** for any four-sided polygon.

Part I: Pattern Block Designs

After the free play, refocus groups on Part I of the activity, creating a group design. As groups work, if they aren't already considering the two questions in Part I, pose these to them.

Groups will likely discover that four squares fit together, three hexagons fit together, and six triangles fit together. Whatever cases they do find, you can point out that these blocks at least *appear* to fit together, but that students can't be sure yet whether they actually fit together perfectly or just come very close. This uncertainty will help foreshadow proving the angle sum formula in *Polygon Angles*.

After approximately 15 minutes of exploration, get students' attention for a brief lecture on **angle**. Angles can be thought of in different ways, and today's activity looks at them from two perspectives. One perspective is dynamic, in which angle is thought of as a turn. The other is static, in which angle is thought of as a geometric figure. For most students, the dynamic concept of an angle as a turn is an easier place to start.

Begin by demonstrating a complete turn. Stand facing the class, make a complete turn, and ask, How far have I turned? You might mention the fact that you have not traveled any distance and therefore the traditional measures of length are inappropriate for measuring a turn. Some students may say that you have turned "one complete turn." Others familiar with degree measurement may say that you have turned 360 degrees. Explain that both answers are correct and that a **degree** is the name for a turn that is $\frac{1}{360}$ of a complete turn. Use the symbol for degrees, writing 360° for the complete turn.

Ask students to demonstrate some other turns. For example, ask everyone to stand and do a half turn. Ask, How many degrees are in that turn? Then have students perform some other fractions of a turn.

Also go from degrees to turns. For example, ask students to turn 120°, and ask the class to describe what fraction of a whole turn that is. Most students need to develop a physical feeling for the turning concept, and this approach lets everyone get involved physically and mentally.

Emphasize that, when in doubt, students can return to the fact that a complete turn is 360°. They can always use this frame of reference to go from a fraction of a turn to degrees, and vice versa.

Ask if anyone knows a special name for a quarter turn. Introduce the term **right angle** and have students figure out how many degrees it must be. Also mention that an angle between 0° and 90° is called an **acute angle** and that an angle between 90° and 180° is called an **obtuse angle**.

Another important way to think about an angle is as a geometric figure (or part of one). To introduce this idea, show students a diagram such as that below, and ask, Where is the angle in this diagram?

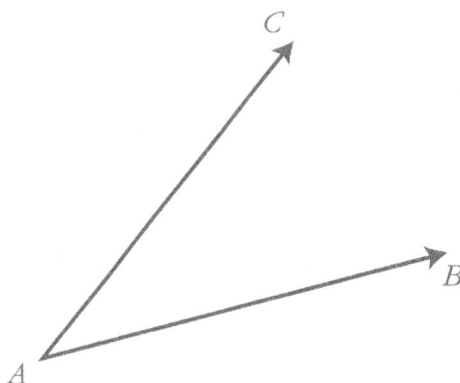

As needed, explain that in order to think of this diagram as showing an angle in the sense of a turn, students can imagine standing at A and facing B, and then imagine turning to face C (while continuing to stand at A).

Extend the lengths of the sides of the angle and ask how this changes the angle itself. Many students confuse side lengths with the size of an angle, so it is important to bring out early and often that the angle itself remains unchanged.

Tell students that point A is called the **vertex** of the angle and that the rays from A through B and from A through C are called the *sides* of the angle. Also introduce the notation ∠BAC, read as "angle BAC." Mention that if there is no chance for confusion, such an angle can be simply referred to as ∠A.

Make it clear that whether we start facing B or facing C, we generally assume that we turn "the short way." Thus, if we start at A, facing B, we would turn counterclockwise to face C, rather than make almost a whole turn clockwise.

Part II: Pattern Block Angles

For Part II, students will need to be familiar with the concept of an angle in a polygon. You can introduce this concept by drawing any polygon. You may need to begin with the terms *side* and *vertex* as applied to a polygon and introduce the plural *vertices* as well.

Then ask students to identify the angles in the polygon. Explain, if needed, that an angle in a polygon is an angle formed where two sides meet at a vertex. Thus, a polygon has the same number of angles as vertices (which is also the same as the number of sides). Use the special case of a square or rectangle to illustrate this fact, and ask students to find the sizes of the figure's angles. They should be able to connect this idea with the earlier discussion and see that each angle is a quarter turn, or 90°.

Have groups now turn their attention to Part II. Explain that they are to determine the measure of the angles in degrees using only the blocks themselves. Remind them to consider what they learned about fitting blocks together to make complete turns.

As students complete Part II, encourage them to continue into Part III.

Once most groups have worked through at least a few of the pattern block angles, bring the class together for a brief discussion on methods and findings. If you have overhead pattern blocks, they will be useful here.

Most of the explanations should be straightforward, such as "I could fit six triangles together at a single point, so each angle is a sixth of a turn, which is 60°." Students will need to do something subtler to find the large angle of the thin parallelogram, such as fit it together with the right angle from a square and an angle from the hexagon. You might again make note that the methods being used to determine these angle measurements are based on the assumption that the blocks fit together perfectly.

Part III: Pattern Block Angles with a Protractor

The protractor is a difficult tool for many students to learn to use. This activity is intended to help students develop a meaningful understanding of and experience with angles so that they can use the protractor to measure angles and get the same results. The activity assumes students will use the known angle measures of the pattern blocks to learn to read that measure on the protractor.

Students will have to learn how to align the vertex of the angle, as well as each side of the angle, with the protractor, and then how to read the measurement. Each protractor works slightly differently, so they will need to spend some time exploring with their own protractors.

Help students begin by suggesting that the protractor is a tool to measure angles. Remind them that they now know the measures of several angles—namely the angles of all the pattern blocks—and explain that they can use that knowledge to investigate how to use the protractor to obtain the same measurements.

Because pattern blocks are rather small, suggest that students trace one angle of a block onto paper and then extend its sides. In the example below (the blue rhombus), the angle is known to measure 60°. Let students learn how to use their protractors to arrive at this measure.

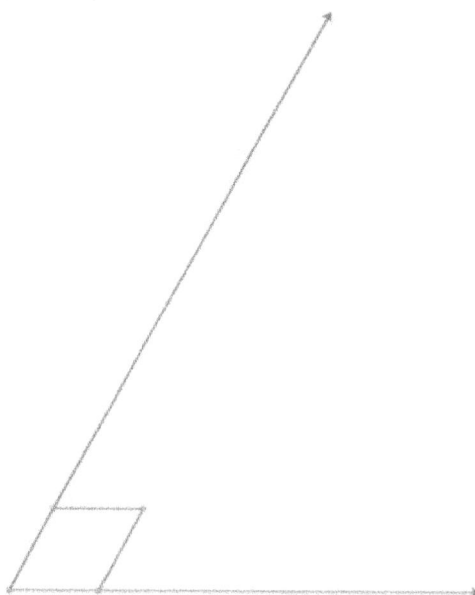

Discussing and Debriefing the Activity

You might conclude the activity with some students demonstrating how to measure angles using a protractor. You can provide more experiences for them to practice this skill, although many opportunities come in the next activities.

Key Questions

How far have I turned?

How many degrees are in that turn?

Supplemental Activities

A Protracted Engagement (reinforcement) is an open-ended activity in which students are asked to decode a message created using angles of different sizes to correspond to different letters of the alphabet, and then to code a message of their own. In the process, they gain additional experience measuring angles with protractors.

From Another Angle (extension) extends students' work with pattern blocks.

Degree Discovery

Intent

In this activity, students explore conjectures about the sum of the angles in triangles and quadrilaterals and gain further practice in the use of protractors.

Mathematics

The central mathematical idea underlying the next three activities—*Degree Discovery, Polygon Angles,* and *An Angular Summary*—is that there is a functional relationship between the number of sides of a polygon and the sum of the measures of its interior angles: sum of angle measures is equal to 180 degrees multiplied by (*number of sides* − 2).

Students get a good deal of practice with measuring angles in *Degree Discovery*. In fact, upon completing this activity, they should have measured at least 20 angles. The design of the activity provides students with feedback on correct protractor use—students tend to check their readings when patterns aren't emerging or when one polygon stands out differently from the others.

Progression

Students will draw several triangles, measure and sum the angles in them, and then do the same for quadrilaterals. Their observations are noted in class and initiate a sequence of activities in which students derive and prove the angle sum formula for polygons. *Degree Discovery* works particularly well as a homework assignment and sets up class work on *Polygon Angles*. *An Angular Summary* serves as a wrap-up assignment for this part of the course.

Approximate Time

20 minutes for activity (at home or in class)

15 minutes for discussion

Classroom Organization

Individuals, then groups, followed by whole-class discussion

Doing the Activity

To transition, mention that in *Pattern Block Investigations*, students found the measures of the interior angles of the special polygons represented by the six pattern blocks. Now they will be asked to make conjectures about the sum of the angle measures of any triangle and then of any quadrilateral.

Point out that students are to draw a variety of triangles. Ask the class to come to an agreement on how many each person should draw (we suggest three at least).

Discussing and Debriefing the Activity

Give students a short time to share results and ask questions within their groups. Then invite some students to share their observations about triangles. Since they are using approximate measurements, their angle sums may not be exactly 180°, but they should see that regardless of the shape of a triangle, the angle measures always seem to add up to about 180°.

This might lead to the conjecture that the angle sum for any triangle is exactly 180°, perhaps using the analysis in *Pattern Block Investigations* for the triangle pattern block as support. But this is a strong statement, one that cannot be proved by measuring, as no measurement is ever exact. What if the real answer were 181° or 179.5°? And what if the result is different for some triangles? This activity is designed to raise, rather than settle, these questions.

A thoughtful argument could arise from the green pattern block triangle. Students should have noted that six of these blocks seem to fit together around a single point, so the angles are apparently 60° each. However, this too is not a conclusive argument, as students have no way yet to be sure that the blocks fit together perfectly.

Use the word **conjecture** to describe the hypothesis that the angle sum for every triangle is 180°.

In the discussion of the next activity, *Polygon Angles*, students will be told that the angle sum for triangles is always 180° and that they will see a proof for this later in the year (in the unit *Shadows*). For now, leave the issue unresolved, so that students are not yet certain whether their conjecture is true.

For Question 3, let students share their conclusions about angle sums for quadrilaterals. They will probably see that the sum always appears to be approximately 360°. Bring out that this observation, like the one for triangles, is only a conjecture (at least for now), since the measurements are only approximations. Have students discuss how their work on *Pattern Block Investigations* relates to this conjecture. Their results for the four quadrilaterals should confirm the conjecture.

Some students might offer that if the angle sum for triangles is 180 degrees, then it makes sense for the sum for a quadrilateral to be 360 degrees, because a quadrilateral can be seen as two non-overlapping triangles. This connection will be explored more deeply in the next activity.

Key Questions

What if your protractor measurements are not exact?

Why might measurement results vary for some triangles?

Polygon Angles

Intent

Building on previous work with patterns, In-Out tables, and functions, this activity asks students to generalize their observations about the relationship between the sum of interior angles and the number of sides of a polygon.

Mathematics

In this activity, students generalize the results from triangles and quadrilaterals to all polygons. This mathematical investigation is a valuable opportunity for them to learn about what doing mathematics is and to see themselves doing mathematics.

Progression

Students work in groups on a rather open task to explore and record what they notice about the sums of polygon angles. If they record their observations about different types of polygons in an In-Out table comparing number of angles to angle sum, they may observe another pattern.

Approximate Time

30 minutes

Classroom Organization

Groups

Doing the Activity

This activity provides another opportunity for students to engage in a fairly open, unstructured exploration allowing them to approach the problem, structure their own time, and organize their data in their own way.

This activity immediately follows the observations made about angle sums for triangles and quadrilaterals. Tell students that now they will explore polygons with more than four sides.

As groups explore, you will have an opportunity to observe who may be having trouble with the protractor. Encourage group members to help each other use the tool properly.

While groups work, encourage them to gather data, make observations, and look for patterns. Ask, **Think about different ways to organize your data to see if there might be patterns in your findings.** By drawing polygons and measuring and adding their angles, students can build an In-Out table of conjectures, like the one below. **What do you notice about your table?**

Number of Sides	Angle Sum
3	180
4	360
5	540
6	720
7	900

Ask questions to help students progress from simple pattern identification to rule building.

Given these results, what might be a conjecture for the angle sum for a 10-sided polygon? A 12-sided polygon? A 100-sided polygon? An *n*-sided polygon?

Is there a general formula for connecting the *In* to the *Out* in this table?

Students may recognize that all angle sums are a multiple of 180 degrees, but what multiple? This generalization—if *n* is the number of sides, then (*n* – 2)180 is the angle sum—is the key underlying functional relationship. It is likely that several groups will notice the generalization, but may not have symbolic notation for the rule; it is not an easy step to recognize that (*n* – 2) can be written to represent "two less than the number of sides." You might encourage students to write their rules as sentences.

The important challenge is the proof that this relationship must always hold, even beyond the data in the table. For groups that have achieved some confidence with this pattern, begin by reminding them it is only a pattern in the shapes they have seen—are they certain the pattern continues?—and then challenge them to prove the relationship they have conjectured.

Why must your rule be true for all polygons?

A slightly simpler question, to get a group started, is, Why should the triangle sum for quadrilaterals be exactly twice that for triangles?

Discussing and Debriefing the Activity

Bring the class together when you think that groups have made good progress exploring and observing patterns and a whole-class discussion can help them to move forward.

You may want to begin by asking students to review what they saw yesterday about angle sums for triangles and quadrilaterals. Then let volunteers state their conclusions about angle sums for polygons with more sides. Although their measurements will again be approximate, they will probably come up with conjectures that can be entered into an In-Out table. Take this table as far as students' results lead, and then ask, **Did anyone come up with a general formula expressing the angle sum as a function of the number of sides?**

If you get a clear statement of the generalization, try to determine whether the class sees where the formula came from. If not, you can build up to the formula by asking students to guess what the angle sum would be for polygons with a specific number of sides not covered yet, based on information in the table.

For example, if the table goes up to a 7-sided polygon, ask students to use the data to formulate a conjecture for 10-sided polygons. **What do you think is the sum of the angles in a 10-sided polygon?** They should probably be able to extend the table by adding 180° three times to get additional rows. By now, they will probably have recognized that the *Out* values all seem to be multiples of 180°.

You can follow up with a large numerical case, such as a 100-sided polygon. **What should you multiply 180° by to get the sum of the angles in a 100-sided polygon?** Students should be able to confirm that the necessary factor seems to be found by subtracting 2 from the number of sides.

Add a row to the table to show this formula.

Number of sides	Angle sum
3	180°
4	360°
5	540°
6	720°
7	900°
8	1080°
9	1260°
10	1440°
100	17640°
n	$(n - 2)180°$

Ask if anyone can explain why the angle sum for quadrilaterals should be exactly twice that for triangles. They should be able to see that a diagonal can be constructed to split a quadrilateral into two triangles; this works even for concave quadrilaterals. Without getting into a lot of detail, use that fact to conclude that the angle sum for a quadrilateral is the sum of the angle sums for its two triangles. Emphasize that this argument does not prove that the angle sum for a triangle is 180° or even that every triangle has the same angle sum. It *does* prove that if every triangle has an angle sum of 180°, then every quadrilateral has an angle sum of 360°.

Finally, ask how the argument for quadrilaterals might be used to explain the formula for the general polygon. Here are two approaches students might use.

They may see—using more examples, if needed—that a polygon with n sides can be divided, using diagonals, into $(n - 2)$ triangles. This is easy to see for convex polygons (all angles less than 180°), but can also be done for concave polygons, with slightly more effort.

They may see that a single diagonal can be used to divide an n-sided polygon into a triangle and an $(n - 1)$-sided polygon. This explains why each side added to the polygon increases the angle sum by 180°. (This approach is another example of recursive reasoning; see the discussion of *Diagonally Speaking*.)

Key Questions

Think about different ways to organize your data to see whether there might be patterns in your findings.

What do you notice about your table?

What would be your conjecture for the angle sum for a 10-sided polygon? A 12-sided polygon? A 100-sided polygon?

Is there a general formula connecting the In to the Out in this table?

All angle sums are a multiple of 180 degrees, but what multiple?

What do you think is the sum of the angles in a 10-sided polygon?

What should you multiply 180° by to get the sum of the angles in a 100-sided polygon?

Why must your rule be true for all polygons?

Why should the triangle sum for quadrilaterals be exactly twice that for triangles?

Supplemental Activity

A Proof Gone Bad (reinforcement) asks students to explain the contradictions in another student's proof.

An Angular Summary

Intent

In this activity, students reflect on and apply their knowledge of the relationship of sides and angles in polygons. This activity emphasizes the important mathematical relationships they have worked on recently, including the unproven fact that the sum of the angles in a triangle is 180° and, based on this conjecture, the proven polygon angle sum formula.

Mathematics

This activity draws upon the polygon angle sum formula to introduce the concept of a **regular polygon**, a polygon in which all angles have the same measure and all sides are the same length. Students draw and measure angles using a protractor one more time.

Progression

This activity serves as a wrap-up for the angle and polygon investigation sequence. After recalling and writing about what they know about polygon angles, students solve the angle measures of a regular pentagon and regular octagon and then draw these polygons.

Approximate Time

20 minutes for activity (at home or in class)

10 minutes for discussion

Classroom Organization

Individuals, followed by small groups

Doing the Activity

This activity requires little or no introduction.

Discussing and Debriefing the Activity

You may want to have one or two volunteers read or present their work on Question 1 and then allow the class to add to or correct the material presented.

Be sure to get some explanations for the angle sum formula. For example, these might refer to the idea of dividing a polygon into triangles or to the numerical pattern found in *Polygon Angles.*

For Question 2, ask students to explain their work. Use the opportunity to briefly emphasize the definition and basic properties of a **regular polygon**. Then ask what difficulties students had in drawing the polygons.

In Question 3, students are asked once more to use protractors to draw polygons, reinforcing their grasp of the pattern developed in the previous activities. If they correctly determine the size of each angle in a regular polygon, and measure correctly, their figures should close—that is, the last side of each figure should meet the first side at the final vertex.

Putting It Together

Intent

As the name suggests, *Putting It Together* encourages students to bring to bear all of the mathematical tools and techniques they have been developing throughout the unit, and their developing identity as a learning community, on a group of summary activities.

Mathematics

In these activities, students will use In-Out tables to solve several problems. One activity will introduce two famous, mathematically important patterns that will surface again later in IMP. Others will combine algebraic thinking, and in particular the concept of function, with ideas from geometry. Finally, students will review and summarize their work over the entire unit in a portfolio.

Progression

Squares and Scoops

Another In-Outer

Diagonally Speaking

The Garden Border

Border Varieties

Patterns Portfolio

Squares and Scoops

Intent

This activity challenges students to draw on their work with patterns and their explorations in *Consecutive Sums* to identify general rules for two important patterns.

Mathematics

Question 1 involves consecutive sums starting with 1. Students investigate the relationship between the height of a stack of squares and the number of squares in the stack. The stack of squares is arranged in a triangular pattern, with each row of squares one unit longer than the one above. A 1-high stack contains 1 square, a 2-high stack contains $1 + 2 = 3$ squares, and a 3-high stack contains $1 + 2 + 3 = 6$ squares. The numbers 1, 3, 6, and so on are called the *triangular numbers*. In general, an n-high stack will contain $1 + 2 + 3 + \cdots + n$ squares.

Question 2 involves an analogous idea for consecutive products starting with 1. It poses a combinatorial question: How many ways are there to arrange n scoops of ice cream on a cone? There is 1 way to arrange 1 scoop and 2 ways to arrange 2 scoops. However, there are 6 ways to arrange 3 scoops. To make the problem easier to think about, imagine that each scoop is a different flavor. For 4 scoops, once the first flavor is chosen, we know there are 6 ways to arrange the rest, and with 4 ways to choose the first flavor, there are $4(6) = (3)(2) = 24$ arrangements altogether. In general, there are $n(n - 1)(n - 2) \cdots (2)(1) = n!$ (read "n factorial") ways to arrange n flavors.

Progression

Each question presents the first few rows of an In-Out table. Students are asked to predict the values in subsequent rows and then to generalize the patterns they used to make their predictions.

The activity is particularly appropriate for small-group exploration. The activity *Diagonally Speaking* follows a similar numeric approach and then challenges students to identify *why* the rule they discover must always hold.

Approximate Time

20 minutes

Classroom Organization

Groups

Doing the Activity

Tell students that they will now explore two very important number patterns—patterns that they will see repeatedly, and in surprising places, in their future mathematics work.

In Question 1, students might see a vertical recursive pattern, in which a value in the second column is found from the previous value, and a "zigzag" addition pattern.

Encourage groups to write a general rule for the patterns they find, but allow that they do not necessarily have to be written with symbols alone. Encourage use of words and sentences as well. Some students may recognize that the number of squares in an n-high stack is equal to the sum of the numbers 1 to n. If so, you might remind them of summation notation, which was introduced during *Consecutive Sums.*

None of these patterns is optimal when searching for, say, row 40, or for row n. In these cases, a rule that relates the *In* value to the *Out* value is best.

For Question 2, students might notice an analogous zigzag pattern, in this case a multiplicative one. If groups focus on the recursive pattern and you decide to challenge them to identify a functional pattern, you might turn their attention to Question 2c. In their solutions, even if they begin with their answer for ten scoops, they will probably say something like, "Multiply this by 11, then by 12, then by 13, and so on, all the way up to 100."

If students recognize some connections, you might remind them of **factorial** notation, mentioned briefly in *1-2-3-4 Puzzle.* The notation for products is analogous to summation notation, using the uppercase Greek letter pi (Π) in place of sigma for sums. For example,

$$\prod_{t=3}^{7} t$$

means $3 \bullet 4 \bullet 5 \bullet 6 \bullet 7$.

Discussing and Debriefing the Activity

There is no need for a formal debriefing of this activity. You might invite posting of solutions or some class discussion if students wish to examine other groups' work.

Depending on students' interest, you might identify the common name for the set of numbers in the *Out* column of Question 1: the *triangular numbers.*

Also of note, the patterns in these In-Out tables would be much harder to find if the entries were not arranged sequentially. You might mention that while this is a good principle for analyzing information, the entries of an In-Out table do not, in general, have to be arranged in any particular order.

Key Questions

Is 1.5 an appropriate input for either of these tables?

What do you call the set of possible inputs for an In-Out table?

What other examples have you seen in which only certain inputs were allowed?

Supplemental Activity

From One to N (extension) asks students to find a simple expression in terms of *n* that allows one to find a sum without repeated addition. If students find such an expression, they look for a proof that their answer is correct.

Another In-Outer

Intent

In this activity, students practice integer arithmetic and finding and using rules for In-Out tables. They also return to the focus on language and symbolic notation begun in the earlier activities *Inside Out* and *Pulling Out Rules*.

Mathematics

The six questions in this activity give students additional opportunities to express the relationships between the *In* and the *Out* in an In-Out table representation of a function. Students write algebraic equations for expressing the *Out* as a function of the *In* and use their rules to find both the *Out* given the *In* and the *In* given the *Out.*

Asking students, in effect, to find both *y* given *x* and *x* given *y* emphasizes the "doing and undoing" aspect of algebraic thinking. The values in these tables also offer students the chance to use their knowledge of integer arithmetic, stressed earlier in the unit in the "hot and cold cubes" activities.

Progression

This activity is particularly appropriate for students to begin as a homework assignment.

Approximate Time

20 minutes for activity (at home or in class)

30 minutes for discussion

Classroom Organization

Individuals, then groups, followed by whole-class discussion

Doing the Activity

Tell students that in this activity, they will look for patterns in more In-Out tables. Some of the tables are tricky and will draw upon their creativity. Students will be asked to write some of their rules as algebraic equations.

Discussing and Debriefing the Activity

Have students convene in groups to briefly share findings and ask each other questions. You might then have the members of various groups begin the discussion by sharing what they noticed in one of the In-Out tables.

Use Questions 2, 3, and 6 to review the usage of the term **function**. Ask volunteers to talk about how they translated their rules from verbal form into algebraic equations.

Questions 1, 4, and 5 are nonnumeric logic puzzles. The most engaging question in the activity is probably Question 5 (the one with the funny faces). If students are stuck, you might suggest making an In-Out table in which the number of eyes and the number of hairs are both *Ins*.

In		Out
Number of eyes	**Number of hairs**	
2	2	6
2	3	11
3	4	19
2	5	?

Finding a rule for this table could be left open. Here are two.

Out = 3(number of eyes) + 5(number of hairs) – 10

Out = number of eyes + (number of hairs)^2

Both formulas fit all three given rows, but the first formula gives 21 for the missing entry, while the second gives 27.

Diagonally Speaking

Intent

This activity revisits and pulls together ideas in the *Patterns* unit. It reinforces In-Out tables and the search for patterns as powerful problem-solving tools. Students use an In-Out table in a geometric context to find a functional relationship and are challenged to prove why their pattern holds.

Mathematics

Students explore geometric ideas in this activity, such as the definition of a polygon diagonal and the connection between the number of sides and the number of vertices of a polygon. They use their developing algebra skills to analyze a geometric situation: the number of **diagonals** in any polygon is a function of the number of sides (or vertices) of the polygon. And finally, returning to the developing notion of proof, students utilize the geometric properties of polygon angles to prove that their numeric pattern must hold for all polygons.

Progression

Students gather and organize data, search for patterns and use them to make predictions, and use the problem context to explain why the patterns they found hold. This activity can be used as a small-group activity or assigned for homework and then discussed in class.

Approximate Time

20 minutes for introduction

20 minutes for activity (at home or in class)

20 minutes for discussion

Classroom Organization

Individuals or small groups, followed by whole-class discussion

Doing the Activity

This activity illustrates a common use of tables in mathematics—to organize information about a complex situation in order to gain insight into the situation itself. The context determines a well-defined, unique function. Part of students' task is to examine how to use that context to justify and explain any pattern or rule they find.

Arouse students' curiosity by posing the simple question that forms the basis for this activity.

Can you predict how many diagonals a polygon has?

Encourage everyone to draw a polygon (suggest from four to seven sides) and count its diagonals. Have a few volunteers report their findings, and then turn students to their groups and encourage them to continue their investigation.

Discussing and Debriefing the Activity

You might begin by drawing a table on the board and calling on someone to fill in one row of the table based on that group's work. Pause after each row and ask students if they agree or if they have any concerns about the numbers just recorded. After several rows have been added, the table will look something like this.

Number of sides	Number of diagonals
3	0
4	2
5	5
6	9
7	14

A *convex polygon* is defined as one in which all diagonals are inside the figure. A *concave polygon* is one in which all diagonals are not inside the figure. Some students might notice that for a concave polygon, one might argue that a segment should not be considered a diagonal if it goes outside the figure. If so, assure them that such a segment does not violate the definition of a *diagonal:* a line segment that connects two vertices of a polygon and is not one of its sides.

Ask for volunteers to describe patterns they found. Encourage them to draw on the board to aid their descriptions. During the discussion, draw out the rule that defines the pattern, and press students to develop a justification for why that rule must always hold. Some examples of patterns, rules, and possible justifications follow.

If students organize the table with increasing inputs, they will more easily notice that the number of diagonals increases by 2, then by 3, then by 4, and so forth, as the number of sides goes up by 1. This vertical or recursive pattern will allow many students to predict that a polygon with 12 sides has 54 diagonals and given that a 20-sided polygon has 170 diagonals, a 21-sided polygon must have 189 diagonals—results they would have great trouble obtaining by drawing the figures and counting diagonals.

The recursive pattern can be summarized in several ways. Two equations that express the relationship are these.

$$Out = previous\ Out + (previous\ In - 1)$$

$$Out = previous\ Out + (In - 2)$$

Some students, drawing on their work in *Consecutive Sums*, might see the following pattern in their tables and predict, correctly, that the number of diagonals in a 12-sided polygon is given by the sum 2 + 3 + ⋯ + 10.

Number of sides	Number of diagonals
3	0
4	2
5	5 = 2 + 3
6	9 = 2 + 3 +4
7	14 = 2 + 3 + 4 + 5

If so, you might ask, **How can you use summation notation to express this pattern of consecutive sums?** The key is determining how to use the *In.* Each *Out* is the sum of whole numbers from 2 to *In* − 2.

Some students might use their previous experience with In-Out tables to try to find a way to relate each *Out* to its corresponding *In* and notice that the following pattern emerges.

Number of sides	Number of diagonals
3	0 = 3(0)
4	2 = 4(0.5)
5	5 = 5(1)
6	9 = 6(1.5)
7	14 = 7(2)

From this, they might see that the multipliers in parentheses are one-half the quantity (*In* − 3), which leads to this closed-formula rule.

$$Out = \frac{In\left(In - 3\right)}{2}$$

Why does this rule make sense? The figure itself provides a clue. If students focus on a single vertex, they will see that the number of diagonals from that vertex is 3 fewer than the number of sides, because no diagonal is drawn to that vertex or to the two vertices immediately adjacent to it.

Key Questions

Can you predict how many diagonals a polygon has?

Can you use the table to predict the number of diagonals for an 8-sided polygon, without drawing one?

Why must all 5-sided polygons have the same number of diagonals?

Why would a 7-sided polygon have five diagonals more than a 6-sided polygon? Why would a 12-sided polygon have ten diagonals more than an 11-sided polygon?

Why does your rule make sense?

Supplemental Activity

Diagonals Illuminated (extension) is a follow-up activity that draws a distinction between recursive and **closed-formula** rules and asks students to develop a closed-formula rule for the number of diagonals of any polygon and to explain why it makes sense.

The Garden Border

Intent

This activity sets the stage for *Border Varieties,* in which students will gain additional experience using algebraic language and symbols to represent geometric situations. In addition, it strengthens students' understanding of equivalent expressions and their skill in working with the distributive property.

Mathematics

The algebra-geometry connection is again a key mathematical element of this activity and the next. In this activity, students derive general approaches for counting the tiles along the border of a square garden. The number of tiles is a linear function of the size of the garden.

Progression

Students begin this two-activity set by creating as many ways as they can think of to count border tiles for a 10-by-10 garden, without counting one tile at a time. They discuss and compare the variety of counting methods they find, setting the stage for developing a symbolic representation of each method and for recognizing the equivalence of the expressions created.

Approximate Time

25 minutes

Classroom Organization

Individuals or groups, followed by whole-class discussion

Materials

Overhead tiles (optional)

Square tiles

Doing the Activity

You may want to present this activity without referring to the student book. If students open their books to *The Garden Border,* they might notice the page for *Border Varieties,* which includes diagrams that provide possible answers for today's activity, and you want students to come up with these ideas on their own.

Introduce the context of the problem, noting that the garden, including the tiles, is to be 10 feet by 10 feet. Draw or display a picture of the tiles, or ask groups to agree what the tile arrangement must look like and have a volunteer share a sketch.

Tell students that their challenge is to figure out how many tiles Leslie needed, without counting the tiles individually, and to write down as many ways as they can for doing this. Then ask that they draw a diagram that indicates how each method works.

Note that you have posed the entire activity verbally. If some students would benefit from written instructions, have them turn to their own books. Or, you could display the instructions for students to refer to.

Discussing and Debriefing the Activity

Ask group representatives to report on one method for calculating the number of border tiles, giving the details of the arithmetic as well as the diagram.

One method is to start with the ten tiles along each edge and subtract 4 to account for the fact that each corner tile is on two edges. For this method, the arithmetic might be 4(10) – 4. Various diagrams can be created to represent this method. The diagram below shows the four 10s along the edges and indicates the four corner tiles that have been counted twice. (A more schematic diagram and five additional approaches appear in *Border Varieties*.)

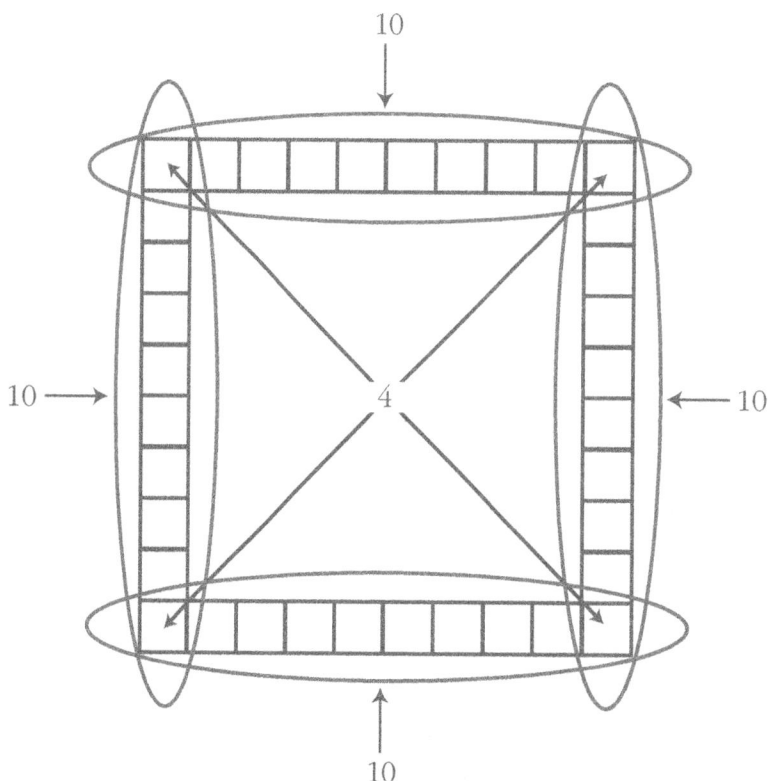

Once the first presenter from every group has reported, ask for any additional ways not yet mentioned. Collect all the approaches that students have come up with.

Take a few minutes to talk about how some of the strategies collected might be applied to a 5-by-5 garden. For example, applying the method described above would show that such a garden uses 4 • 5 – 4 = 16 tiles.

Border Varieties

Intent

This activity gives students additional experience using algebraic language and symbols to represent geometric situations. It also strengthens their understanding of equivalent expressions and skill in working with the distributive property.

Mathematics

In *The Garden Border,* students perceived a geometric context in many ways. The variety of approaches they developed lead to different-looking, but equivalent, methods for counting tiles. Here are three such methods for counting border tiles, along with rules that reflect these three ways of viewing the problem.

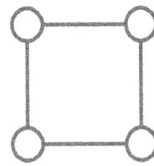

$4(s - 1)$ $2(s) + 2(s - 1)$ $4(s - 2) + 4$

These rules are **equivalent** because each gives the same total number of tiles for a given value of s.

Progression

In this follow-up activity to *The Garden Border,* students compare a variety of methods for counting border tiles for a 10-by-10 garden and then review and generalize these methods for gardens of any size. The classroom conversation emphasizes the equivalence of the resulting expressions and the occurrence of the distributive property.

Approximate Time

5 minutes for introduction

20 minutes for activity (in class or at home)

20 to 40 minutes for discussion

Classroom Organization

Individuals, followed by whole-class discussion

Doing the Activity

This activity asks students to create "formulas" and offers the example $4s - 4$. In the early use of symbolic algebra in IMP, words like *formula* and *rule* are not precisely defined. In the context of this activity, students will be comparing expressions, so an equation is not necessary.

Have students read up to Question 1, and then ask a volunteer to summarize what was stated. Invite others to add points of clarification. Once the important points have been covered, tell students that they will be creating similar formulas for several more methods.

This might be a good assignment to collect to help assess and support students' understanding of using variables to express generalizations.

Discussing and Debriefing the Activity

Ask students to share their answers. They will probably want to use the diagrams in the activity to explain how they found their formulas. The diagrams will help make the generalizations more understandable.

To help clarify the techniques, ask students to test another specific case, such as $s = 100$. **Would this method work for a 100-by-100 square?** For example, for Question 1, the diagram suggests that the border for a 100-by-100 garden would have 100 tiles along the top and along the bottom, leaving 98 for each of the other two sides. Getting students to express this as $2 \cdot 100 + (100 - 2)$ will help elicit the expression $2s + 2(s - 2)$. You might use the phrase "imitating the arithmetic" to describe this technique for developing algebraic formulas to represent situations. Students might also express the method in Question 1 with the formula $s + s + (s - 2) + (s - 2)$.

The various expressions students develop provide an excellent opportunity to review equivalent expressions, the distributive property, and the idea of combining terms.

As much as possible, have students demonstrate equivalence for all the formulas they found for the various diagrams. For example, for Question 1, you might ask a volunteer to explain how to be sure that the expressions $s + s + (s - 2) + (s - 2)$ and $2s + 2(s - 2)$ are equivalent, independent of the problem setting. **How can you be sure these expressions are equivalent?** Students might recognize that $s + s$ is the same as $2s$ and that $(s - 2) + (s - 2)$ is the same as $2(s - 2)$. They might also see why these expressions are equivalent to $4(s - 2) + 4$.

At any time during this discussion, you might use numeric examples to check equivalence or to simply help students see that the operations being carried out in the different orders do yield equal values. Communicate that such examples do not prove equivalence, just help to confirm it.

To wrap up the conversation, remind students of the distributive property and ask them to identify instances in which they determined that two expressions were equivalent and in which they can see the distributive property in action, such as $2(2s - 2) = 4s - 4$. You might ask groups to work on this question for a short period of time and then to share with the class any other instances they found. Record their observations as equalities; for example, $4(s - 1) = 4s - 4$ and

$2(s - 1) = 2s - 2$. Students will likely see that the distributive property is just a subset of the bigger idea of equivalent expressions.

A general statement of the distributive property might look like this.

The expressions $N(a + b)$ and $Na + Nb$ are equivalent.

You might want to post this statement, or another that students develop, for students to refer to.

Key Questions

Would your method work for a 100-by-100 square?

How can you be sure these expressions are equivalent?

Supplemental Activities

More About Borders (extension) contains variations on the *Border Varieties* activity.

Programming Borders (extension) asks students to write a program that answers some or all of the questions posed in *More About Borders*.

Patterns Portfolio

Intent

Students review and document their mathematical activity and learning during the course of the unit. Their product is an opportunity for assessing what they have learned and what they believe is important in their learning (see "About Portfolios" in the Overview to the Interactive Mathematics Program).

Mathematics

Students review their work on all the mathematical topics of the unit, with an emphasis on In-Out tables, proof, and integer arithmetic. Other ideas, such as functions and the distributive property, are approached more informally at this stage and are not specifically addressed.

This metacognitive activity is an important mathematical task in itself, in that mathematics learned in school has a quality distinct from the natural mathematical activity that emerges from interaction with real problems, or at least problems that are real to students. The distinction involves learning the agreed-upon conventions, notation, and terminology for referring to particular ideas that the formal study of mathematics requires.

Progression

Students are formally introduced to the notion of a portfolio and the particular expectations for the *Patterns* portfolio. They review their materials and begin to compile their portfolios as they write their cover letters.

Approximate Time

20 minutes for introduction

30 minutes for activity (at home or in class)

Classroom Organization

Whole-class introduction, then individuals

Doing the Activity

Prior to this day, emphasize the need for students to bring all their work from the unit to school. For some students, having to assemble a portfolio can become a reminder for the need to maintain some organization of their materials.

Disorganization can be an early indicator of students who may find less success in school. You might invite students who are particularly disorganized to visit with you privately to help them organize their materials from the unit, possibly into a three-ring notebook, and then draw out particular activities for their portfolios. Such private time will give you an opportunity to make a personal connection with students as well as support them in getting off to a good start in your class.

Before students begin work on their portfolios, you may want to ask what they recall about portfolios from the discussion at the beginning of the unit.

What is the purpose of a portfolio?

What would be good items to include in a portfolio?

After this brief review of portfolios in general, ask students to read *Patterns Portfolio* carefully. You may then wish to lead a general review discussion of the unit before students begin assembling their portfolios, or you may prefer to let students work in groups or on their own.

The primary goal of this first IMP portfolio is that students have some success becoming aware of and completing each of the three phases: Cover Letter, Compiling Papers, and Personal Growth. The writing of cover letters and the selection of portfolio materials are intertwined activities. Students should probably at least begin their general review of the unit before selecting portfolio materials, but they will need to make certain selections in order to complete the cover letter.

You might suggest two possible orders for work on the portfolios. Students could (1) identify papers to include, (2) write their cover letters, and (3) write their personal growth statements. Or they might (1) write their cover letters only through the point of describing the central ideas of the unit, (2) identify papers to include, (3) complete the cover letters by explaining their choices, and (4) write their personal growth statements.

As students review the unit and their work, they should think about both the mathematical content of the unit and the quality of their writing. Suggest that they choose work that conveys the essence of the unit as well as work that illustrates their expertise in solving and writing about problems.

Have students complete the *Patterns* portfolio for homework. Ask that they bring the portfolios back tomorrow with the cover letter as the first item. You might also mention that they will be able to refer to their portfolios when they work on unit assessments.

Discussing and Debriefing the Activity

You might have students share their portfolios as part of the closing activity to the unit, the *Unit Reflection,* by reading cover letters and/or personal growth statements as a class or in small groups, sharing portfolios among group members, or having a class conversation about what was learned during the unit.

The portfolio is a very personal product for many students, something they may take a great deal of pride in. Public recognition of the thoughtfulness that students invested in creating their portfolios will add to the classroom learning environment and to students' connection to that environment.

Key Questions

What are portfolios? What should go into them?

Lonesome Llama

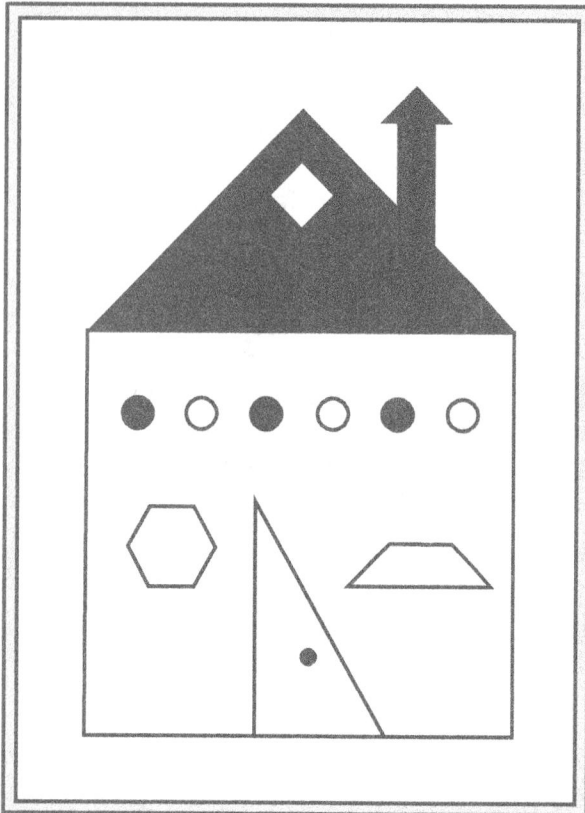

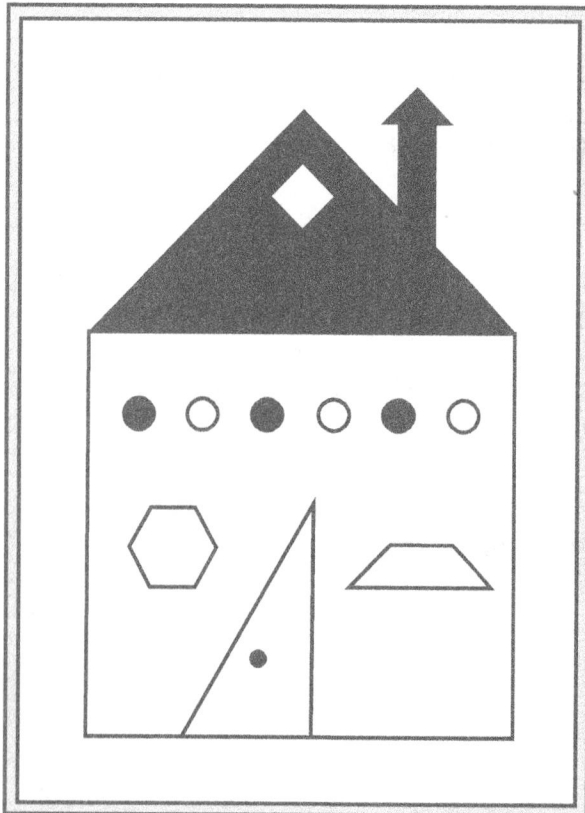

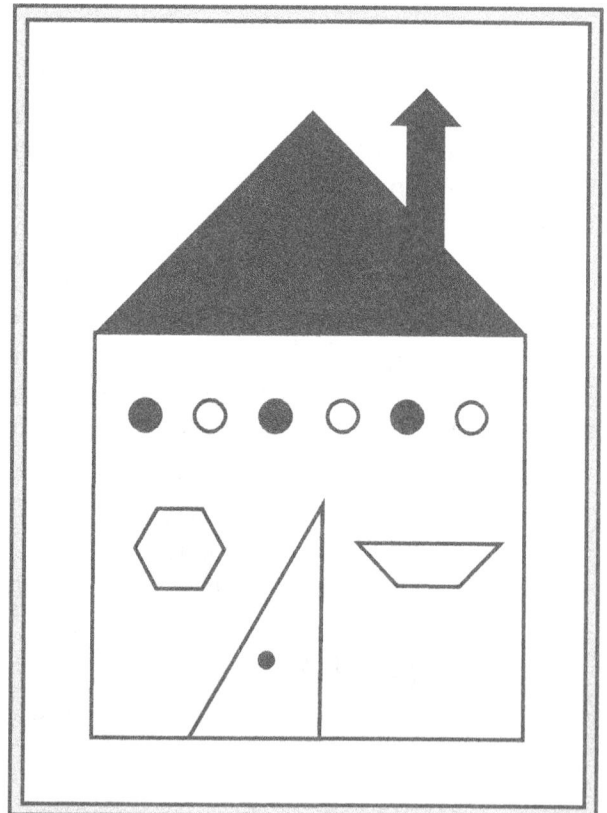

Lonesome Llama (continued)

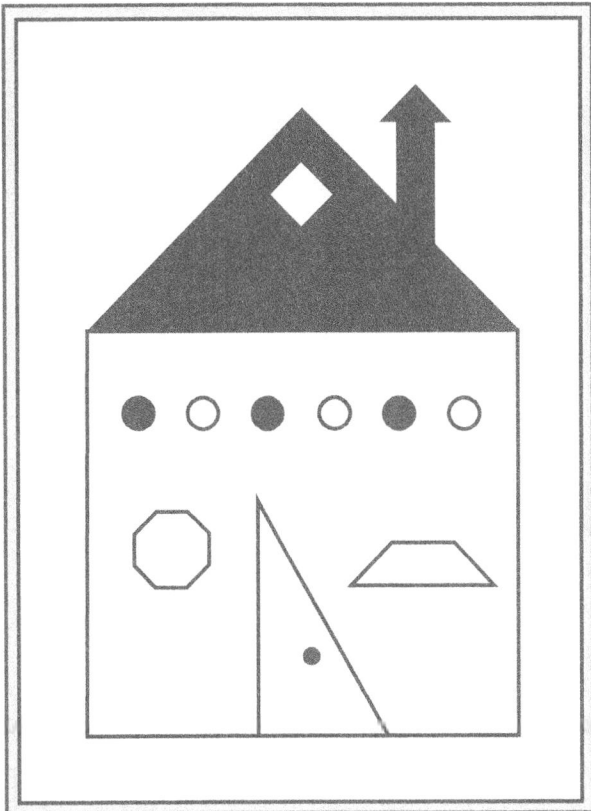

Lonesome Llama (continued)

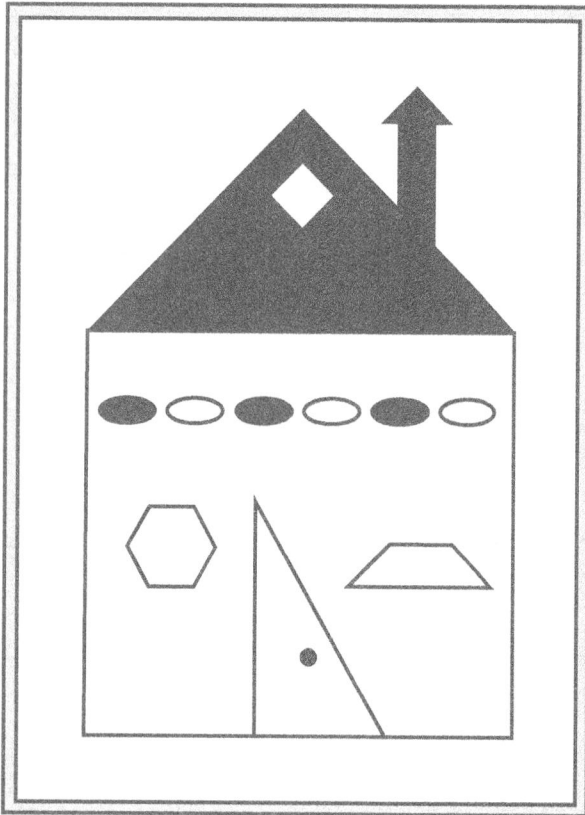

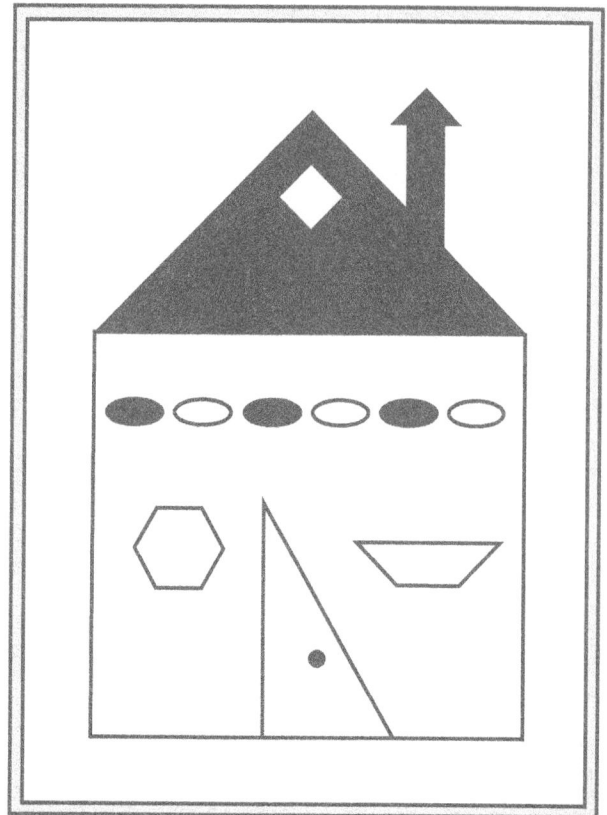

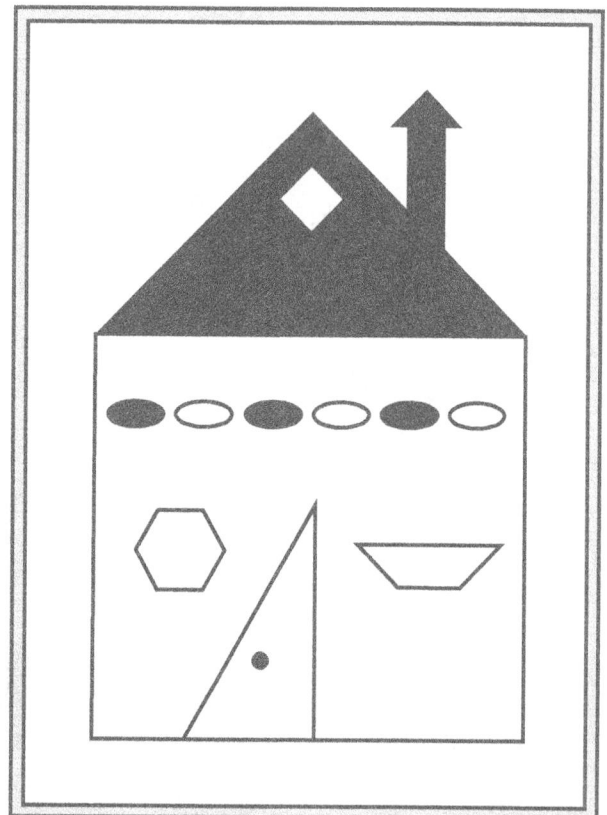

Lonesome Llama (continued)

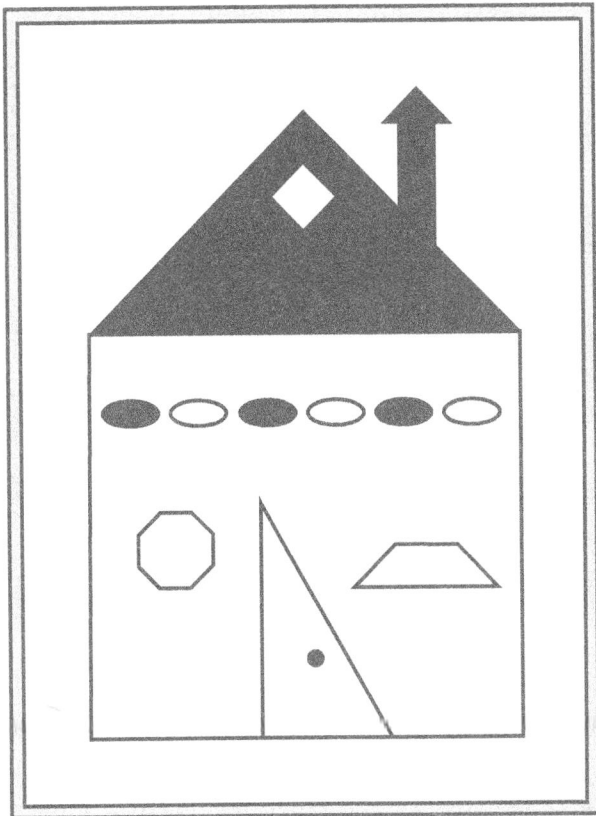

Lonesome Llama (continued)

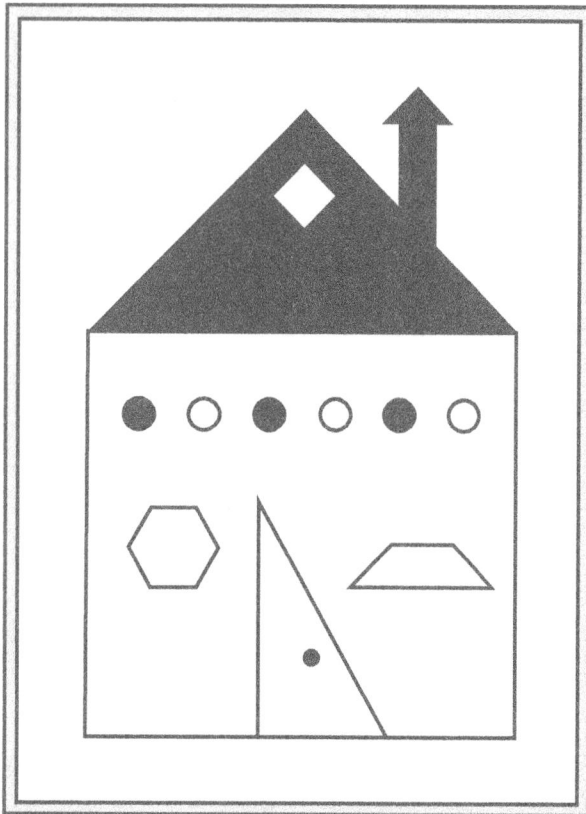

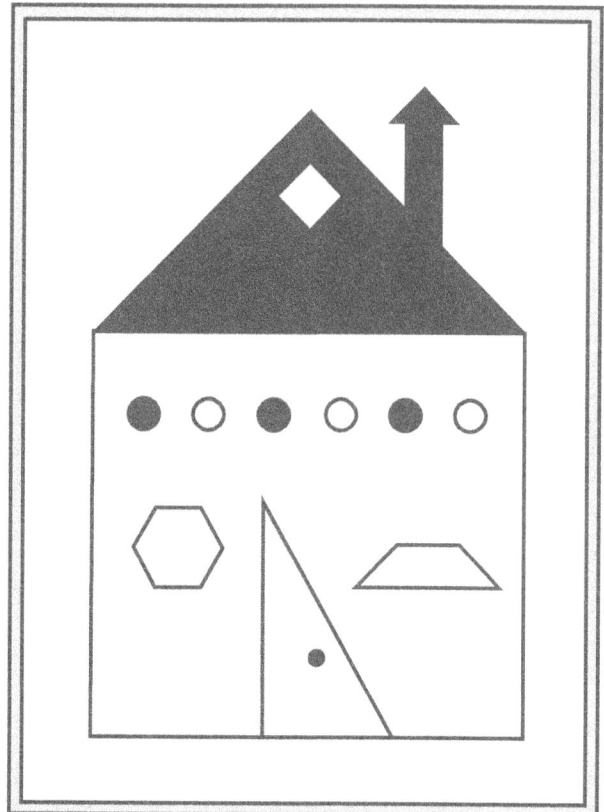

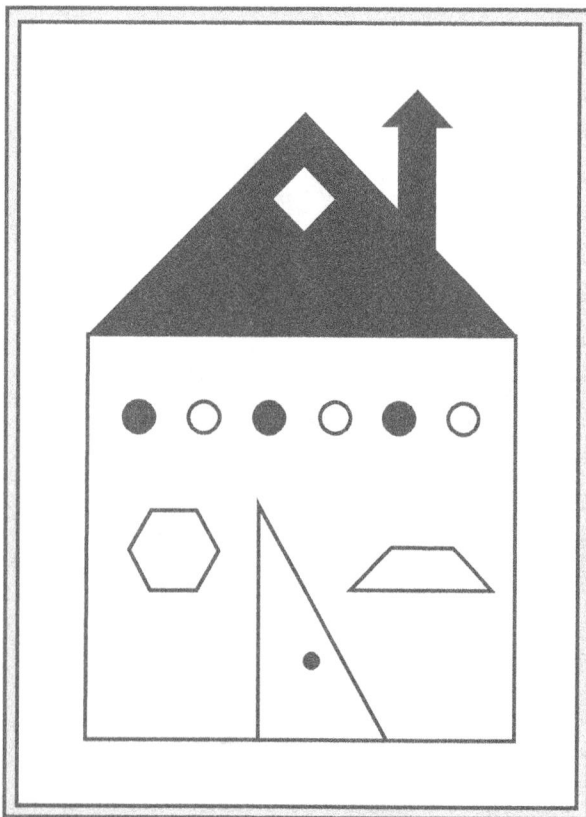

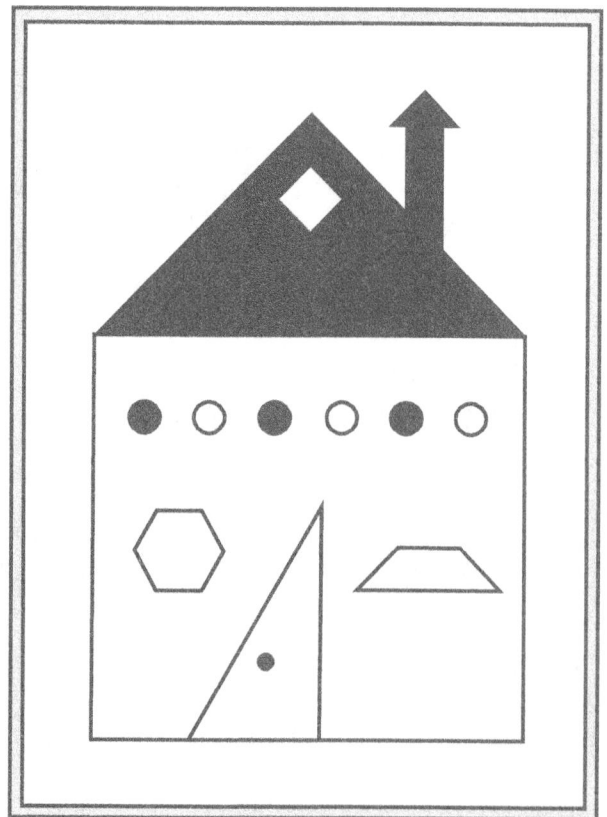

Lonesome Llama (continued)

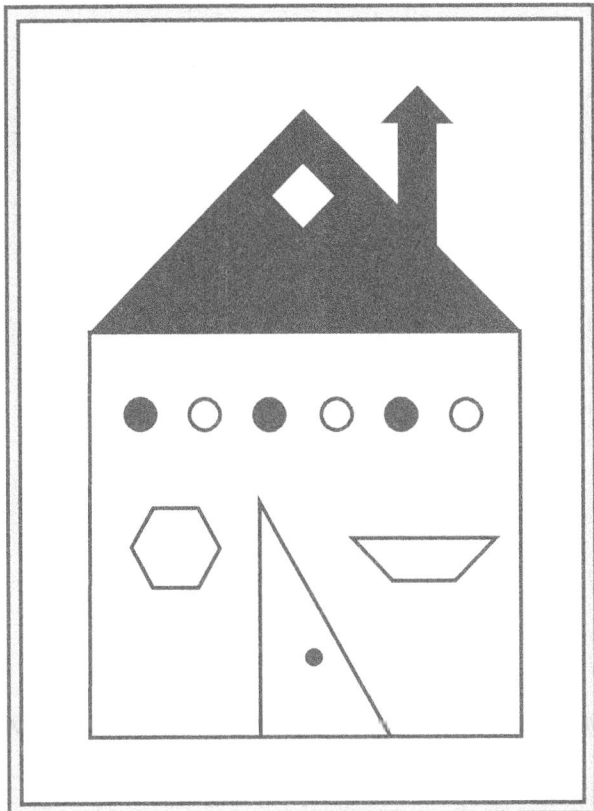

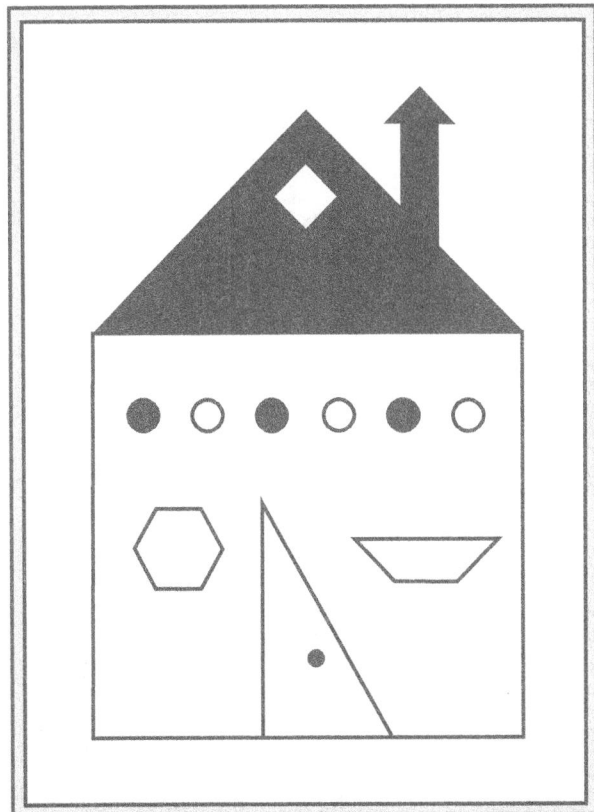

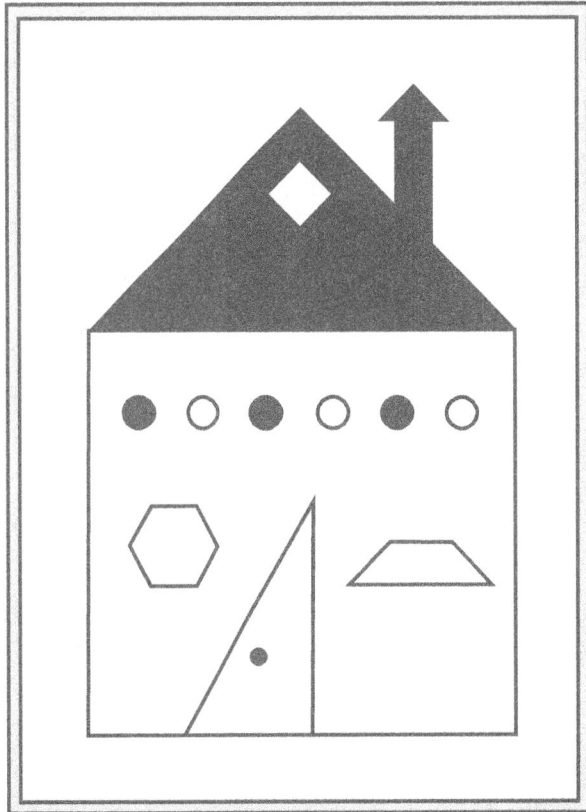

Lonesome Llama (continued)

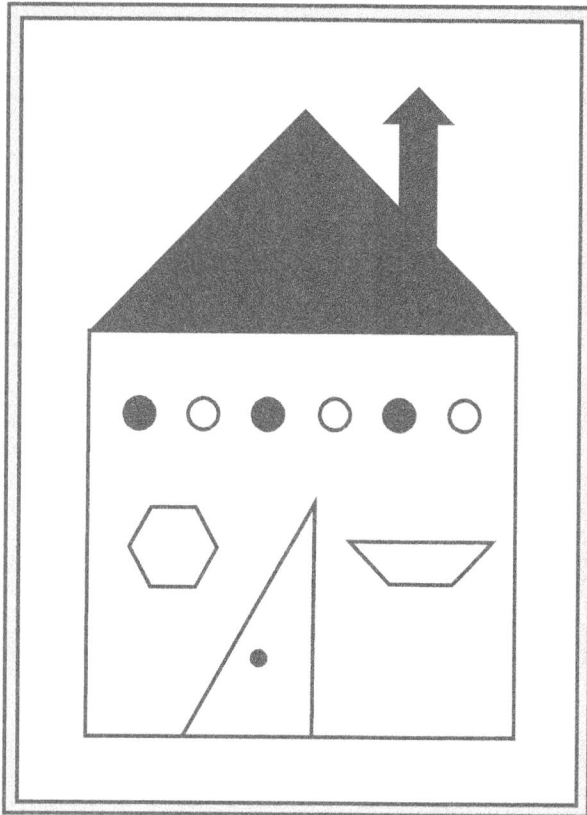

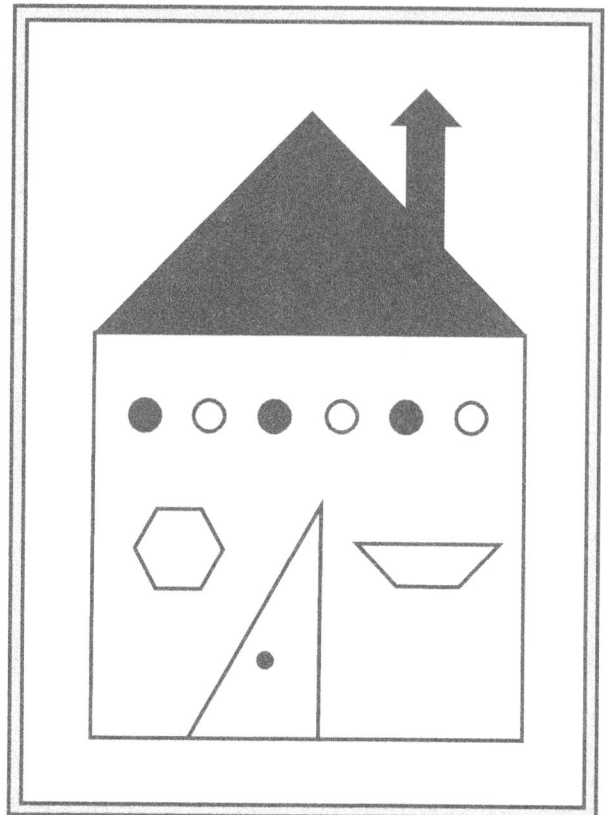

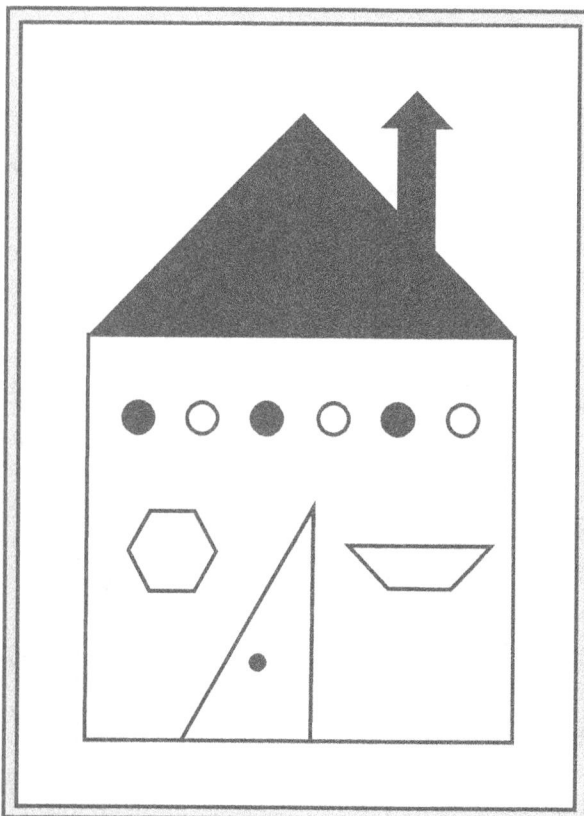

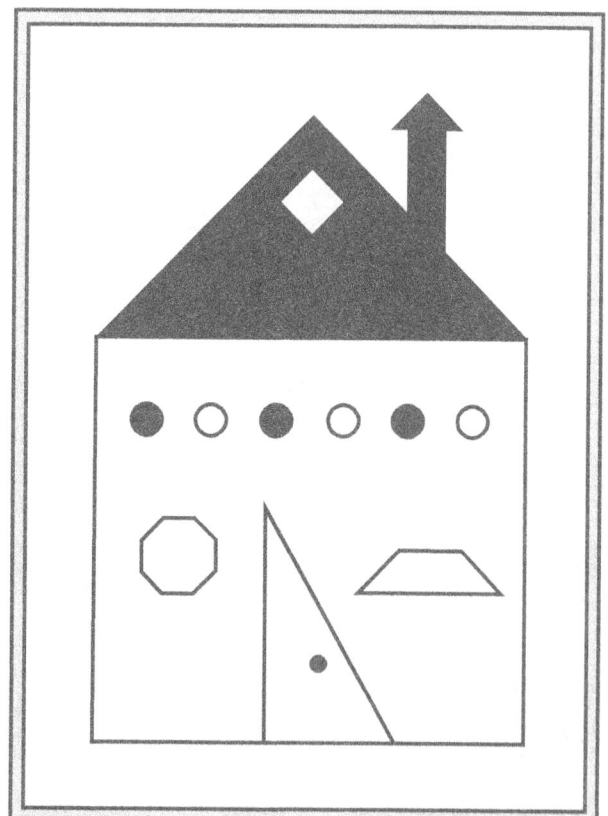

Lonesome Llama (continued)

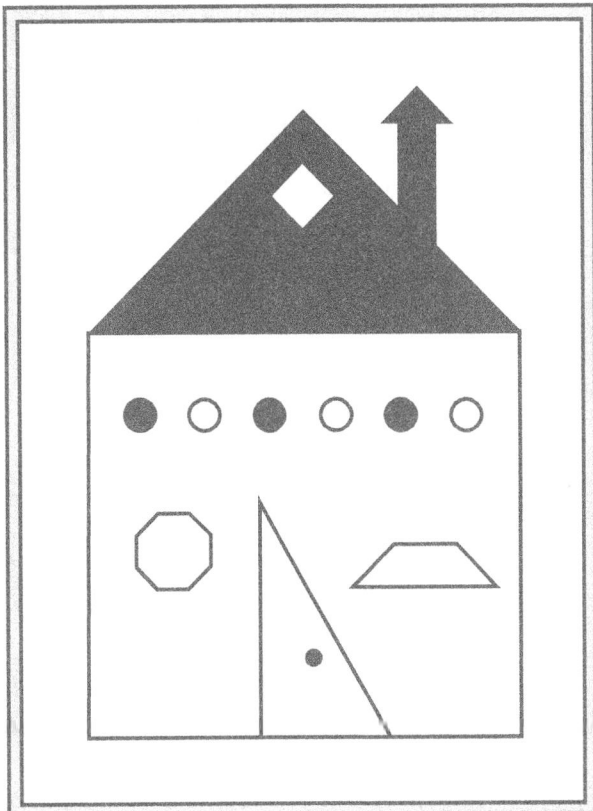

Lonesome Llama (continued)

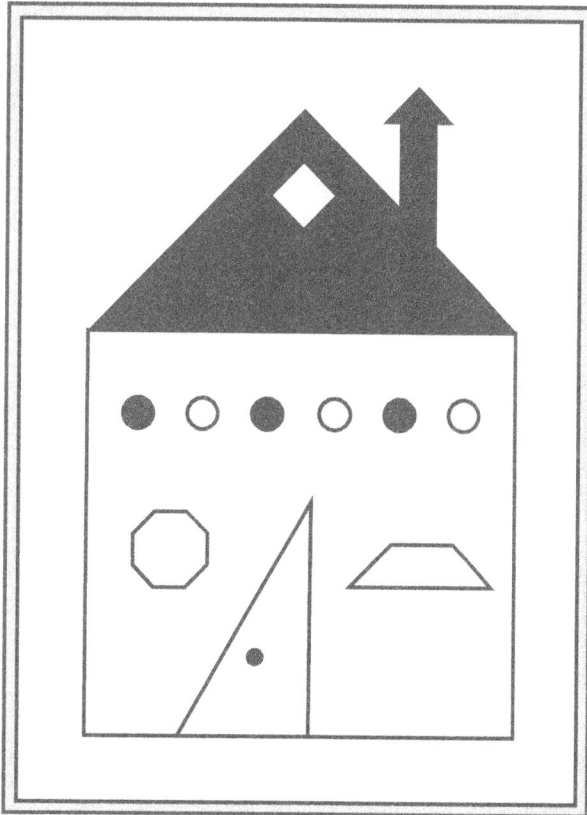

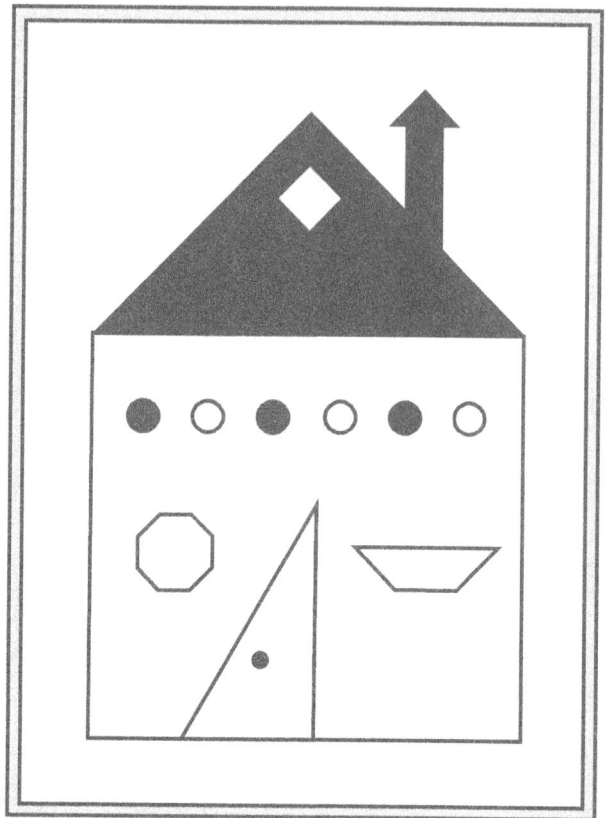

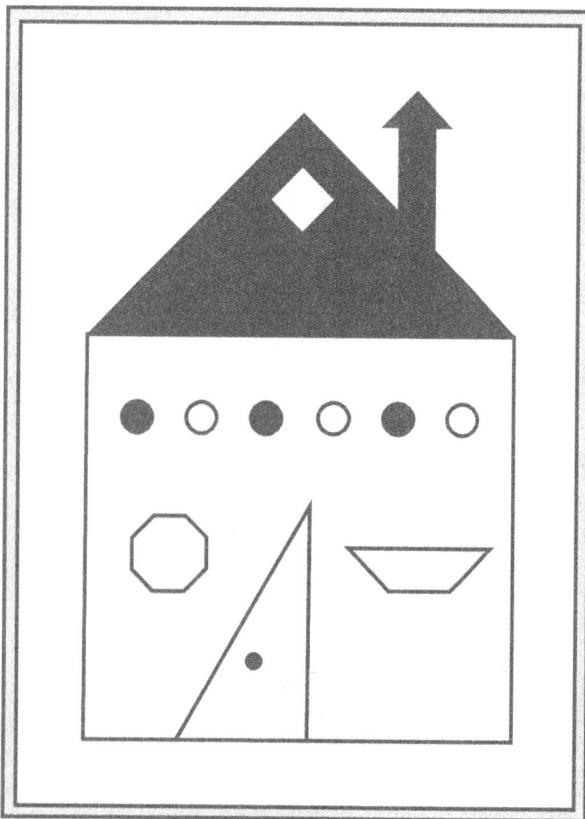

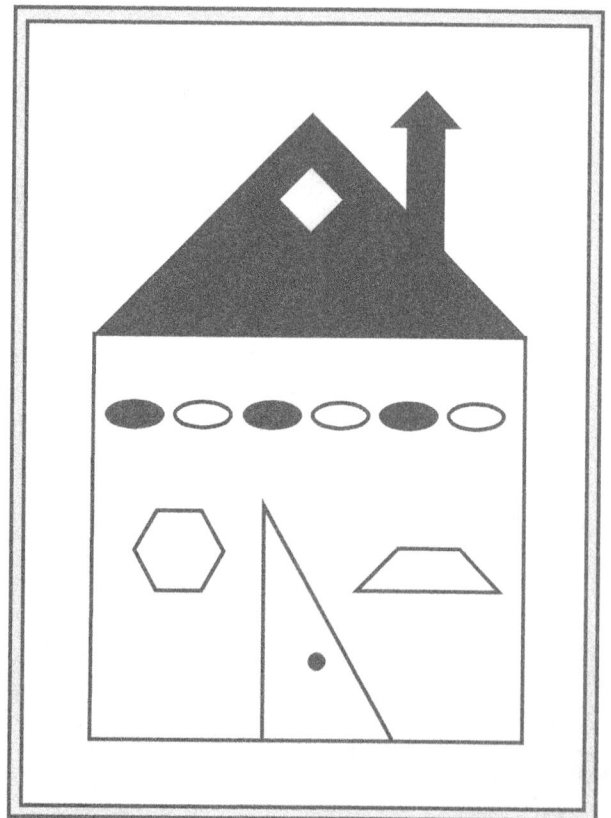

Lonesome Llama (continued)

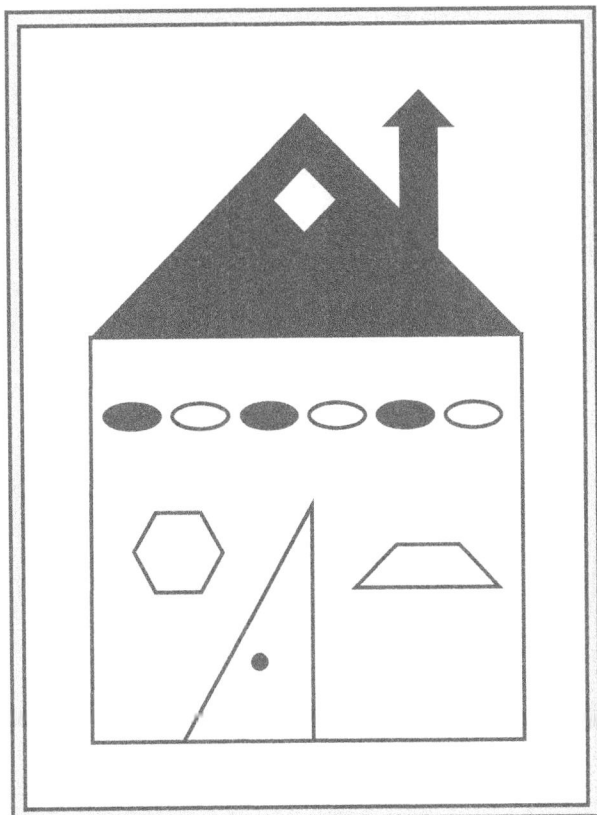

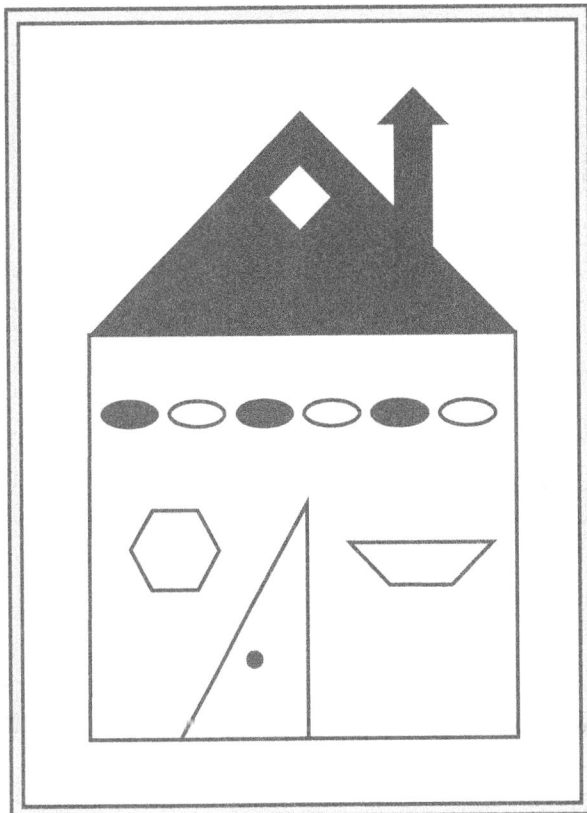

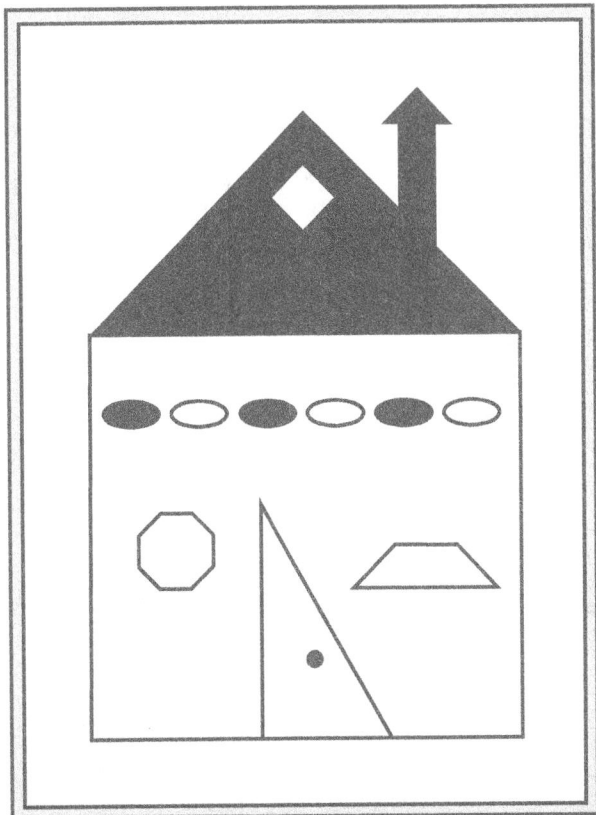

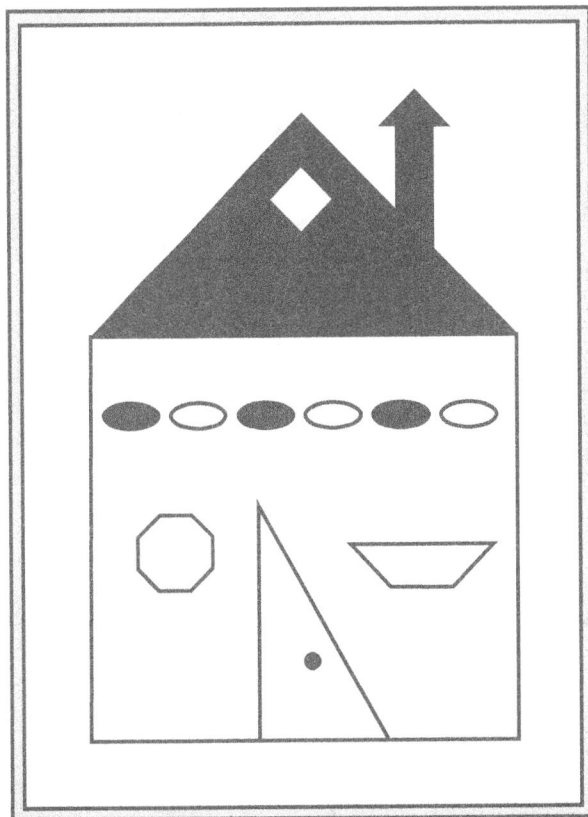

Lonesome Llama (continued)

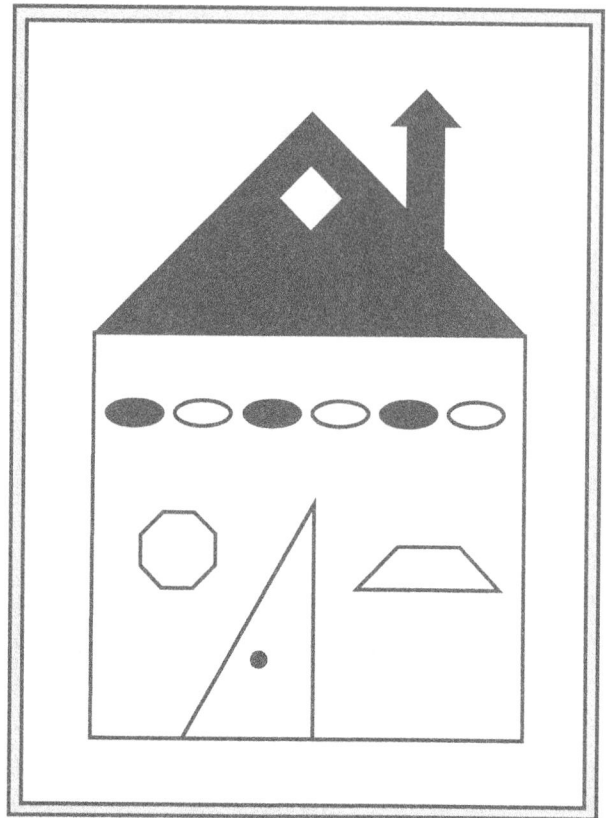

Lonesome Llama (continued)

1-Inch Graph Paper

$\frac{1}{4}$-Inch Graph Paper

1-Centimeter Graph Paper

1. Explain each of the following problems in terms of the model of hot and cold cubes. Each explanation should include a statement of how the temperature changes overall.

   a. $4 - (-3)$

   b. $4 \cdot (-5)$

2. a. Find the missing entries for the In-Out table.

   b. Describe the function represented by the table both in words and by using an algebraic expression.

In	Out
4	7
8	15
-2	-5
10	?
-5	?
?	31

3. An **isosceles triangle** is a triangle in which at least two of the angles are equal.

   The figure at right represents a general isosceles triangle in which $\angle B$ and $\angle C$ are equal.

   a. Suppose $\angle B$ is 60°. Find the size of $\angle A$.

   b. Suppose $\angle B$ is 72°. Find the size of $\angle A$.

   c. Make an In-Out table and develop an expression that will tell you the size of $\angle A$ in terms of the size of $\angle B$. That is, $\angle B$ should be the *In* and $\angle A$ should be the *Out*.

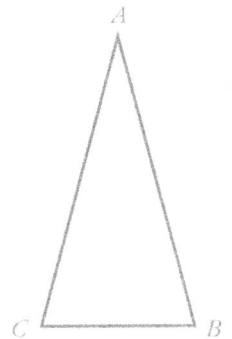

This problem is like the border problem, but it involves cubes instead of squares.

Imagine that you have a rectangular solid with dimensions 5 inches by 5 inches by 7 inches. This large rectangular solid is made up of smaller cubes. Each small cube is 1 inch on every edge.

Someone comes along and paints the large rectangular solid on all of its faces, including the bottom. None of the paint leaks to the inside.

How many of the small cubes have paint on them?

How many have only one face painted?

How many have two faces painted?

Your write-up of this problem should have these components:

• *Problem Statement:* Restate the problem in your own words.

• *Process:* Show all the work you did in solving this problem. If you used tables or diagrams, include them.

• *Solution:* Give your solution and justify it so that someone else will be convinced that your solution is correct.

• *Extra credit:* Suppose the dimensions of the rectangular solid are 5 inches by 5 inches by n inches. Find a formula that gives the number of small cubes with only one face painted.

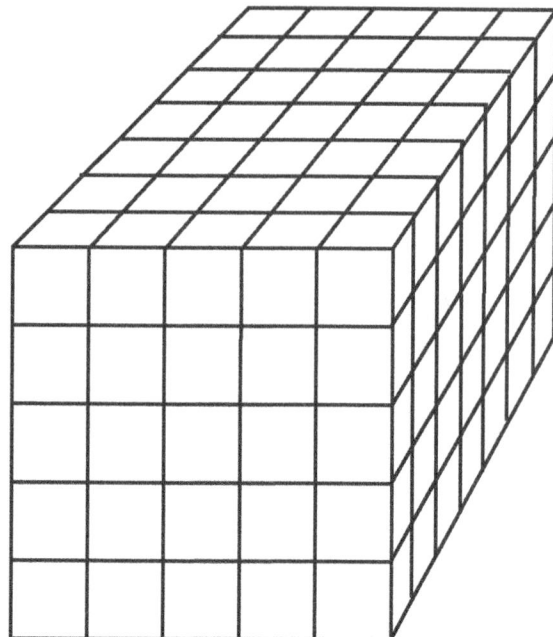

Patterns Guide for the TI-83/84 Family of Calculators

This guide gives suggestions for selected activities of the Year 1 unit *Patterns.* The notes that you download contain specific calculator instructions that you might copy for your students. NOTE: If your students have the TI-Nspire handheld, they can attach the TI-84 Plus Keypad (from Texas Instruments) and use the calculator notes for the TI-83/84.

Students will come in with a wide range of knowledge about graphing calculators and other technologies. The goal is for technology to become a tool that helps students ask and explore interesting questions. Students will have an opportunity to learn new things about graphing calculators during the open-ended *Calculator Exploration.* Until then, keep calculators available and let students discover and teach each other.

For one activity, this guide provides support for required calculator use in the text. For other activities these notes contain ideas about how to use calculators for enrichment or extension of the activities or hints and suggestions for pitfalls to avoid.

POW 1: The Broken Eggs: In their work on the first POW, students may wish to use their calculators to list possibilities for the number of eggs. For example, when they consider the statement "She knows that when she put the eggs in groups of three, there was one egg left over," students might generate the sequence of numbers that match this statement on their calculators: 4, 7, 10, 13, and so on. In calculations that involve performing the same operation (in this case, adding 3) over and over again, the calculator provides a shortcut. This shortcut is described in the *Patterns Notes* in the section "Using the Previous Answer."

Calculator Exploration: In this activity, students explore the capabilities of their graphing calculator without structured directions. Teachers have found that relatively unstructured exploration can help build students' confidence about using technology. Students realize that they can figure out a great deal without asking for directions or help. For this reason, use the notes "TI Calculator Basics" conservatively with students. You might review the material to remind you of the things you will eventually want to bring out in discussions if they are not mentioned by students.

1-2-3-4 Puzzle: Students who use their calculators for this activity will want to find the square root command and the factorial command. They should also know that the calculator uses parentheses exclusively as grouping symbols. (That is, brackets and braces are reserved for other contexts.)

Uncertain Answers: The TI calculator follows the order-of-operations rules. Nevertheless, encourage students to use parentheses in long calculations. Parentheses make calculations easier to follow both on and off the screen.

Extended Bagels: In this activity, students use algebraic expressions to describe the relationship between two columns of a table. You can use algebraic expressions to generate tables like these on the TI calculator. You probably don't want students to work with these tables on a calculator until they are more comfortable with the ideas of tables and algebraic expressions. The instructions in the section "Using Lists to Build a Spreadsheet" describe how to create the table of values containing Marcella's bagel count.

Two calculator features create tables: Table and List. For instructions on using the Table feature, see "Graphing Basics" in *The Overland Trail Notes*. You will notice that the List features allow the calculator to function like a simple spreadsheet.

Do It the Chefs' Way: Students frequently confuse the negative key with the subtraction key. The negative key, which is located beside the decimal key and has a negative in parentheses, changes the sign of the calculation that follows it. The subtraction key, which is located above the addition key, subtracts the next term from the previous term. You cannot begin a calculation with the subtraction key. Use the negative key to represent negative numbers. On the screen, a negative is shorter and raised when compared to subtraction.

Add It Up: It is possible to use 2nd [**LIST**] to build a sequence. It is also possible to sum this sequence. The steps to do this are fairly complicated, and the notation is different from the traditional sigma (Σ) notation, so avoid showing students the calculator procedure right away. However, if your students become comfortable with the activity and you have some extra time, they will probably enjoy seeing how quickly the calculator can sum complicated sequences.

Programming Borders: The note "Programming Your Calculator for Borders" is designed to assist with this activity. It does not address the general concept of a program, so you will still want to discuss that with the class. Instead, it focuses on the mechanics of creating and entering a simple program on a TI calculator.

The *Patterns Notes* contain a simple program BORDER, or BORDER1. There are two modifications: BORDER2 includes some display; BORDER3 behaves the way the supplemental activity in the textbook implies and is advanced, requiring some commands not mentioned in the notes. To save time and prevent transcription errors, you could load the programs BORDER1.8xp, BORDER2.8xp, and BORDER3.8xp in student calculators.

PROGRAM:BORDER

:Input S

:4S–4 → K

:Disp K

PROGRAM:BORDER2

```
:Disp "SIDE LENGTH?"
:Input S
:4S−4→K
:Disp "TILES:",K


PROGRAM:BORDER3
:ClrHome
:Disp "HOW WIDE IS THE"
:Disp "GARDEN?"
:Input W
:Disp "HOW LONG IS THE"
:Disp "GARDEN?"
:Input L
:Disp "HOW WIDE IS THE"
:Disp "BORDER?"
:Input B
:2*B*(W−2B)+2*B*(L−2B)+4*B*B→T
:(W−2B)*(L−2B)→S
:3*T+0.20*S→C
:ClrHome
:Disp "YOU WILL NEED"
:Output(2,1,T)
:Output(2,6,"TILES AND")
:Output(3,1,"WILL HAVE TO")
:Output(4,1,"COVER")
:Output(4,7,S)
:Output(5,1,"SQUARE FEET WITH")
:Output(6,1,"TOPSOIL.")
:Output(7,1,"THIS WILL COST")
:Fix 2
:Output(8,1,C)
:Float
:Output(8,9,"DOLLARS.")
```

Using the Previous Answer

Sometimes you want to perform a sequence of calculations in which each answer builds on the previous one.

For instance, suppose you want to generate the sequence $4, 7, 10, \ldots$, in which you add 3 to generate the next term every time. Press 4 [ENTER] to get the first line of this display, and then + 3 [ENTER] to get the next line. The calculator assumes you are adding onto the previous answer, in this case, 4.

After you perform an operation in this manner, the calculator remembers the operation. Repeat the operation as many times as you wish by successively pressing [ENTER].

You can also use the previous answer in any part of the next calculation by pressing [2nd] [ANS].

```
4
            4
Ans+3
            7
```

```
4
            4
Ans+3
            7
           10
           13
           16
```

TI Calculator Basics

Many of the keys on the TI calculator have two other operations in addition to the primary operation written on the key itself. The second operation appears in small letters above the key on the left. The ALPHA operation appears above the key on the right.

Using the [2nd] Key

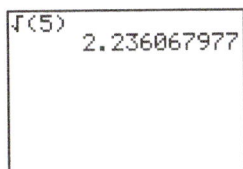

The second operation usually performs an operation that is the inverse, or opposite, of the primary operation on the key itself. For example, the x^2 key has the square-root function as its second operation, because squaring and finding the square root often work as inverses. Notice that the [ON] key has OFF as its second operation.

To indicate a second operation, this calculator guide uses the [2nd] key and then shows the operation itself in brackets. For example, [2nd] [√] indicates the square-root function above the x^2 key.

Now try it yourself. Use the [2nd] [√] key to find the square root of five. The square root function includes an open parenthesis, so you should close the expression within the radical with the close parenthesis. Check your answer with the x^2 key.

Returning to the Home Screen

There are several different displays, or screens on your TI calculator. The main screen is called the home screen. If you find yourself at another screen, press [2nd] [QUIT] to return to the home screen.

Editing What You Type

To clear your home screen, press [CLEAR].

Enter a calculation like the one shown here, but don't press [ENTER] yet. You can use this calculation to practice your editing.

Continued on next page

```
5*1■73
```

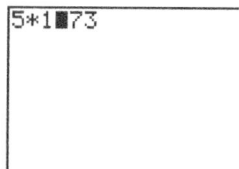

To replace a character or command, use the arrow keys to place the cursor over the character or command you wish to replace. Then type the new character.

To delete an entry, place the cursor over the entry you wish to delete and press [DEL].

To insert a character or command, place the cursor on the character that will follow your new entry and press [2nd] [INS]. Then type what you want to insert.

Recalling a Previous Entry

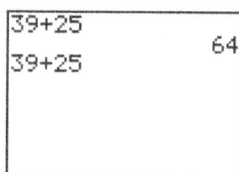

```
39+25
39+25        64
```

To recall a previous entry, press [2nd] [ENTRY]. This will repeat the last entry you typed onto your screen. If you press [2nd] [ENTRY] repeatedly, you will move back through your work one line at a time, so you can actually recover something you did several steps earlier.

Using [ALPHA] and [A-LOCK]

```
ALBERTO■
```

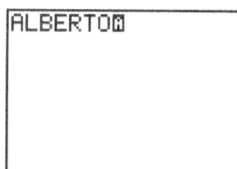

The [ALPHA] key puts you in ALPHA mode, which allows you to use the secondary operations above the keys on the right. Most of these are letters of the alphabet. For example, pressing [ALPHA] [A] produces the letter A on your screen.

The [ALPHA] key only affects the very next key you press. That is, ALPHA mode lasts for only one key stroke. Press [2nd] [A-LOCK] to put your calculator into ALPHA mode so that it stays in this mode. Now write your name on your calculator screen.

Working with Exponents

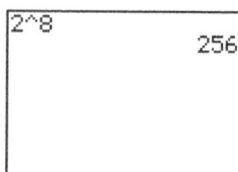

```
2^8
           256
```

We often use the ^ symbol to indicate an exponent. The TI calculator uses this symbol, too. For example, you use the keystrokes [2] [^] [8] to calculate 2^8.

Continued on next page

We raise numbers to the power 2 often enough that the calculator provides the x^2 key for this purpose.

Using the MODE Key

Press MODE. This key gives you the option of controlling certain aspects of how the calculator displays and interprets information. (The screen shown here is a TI-84 Plus screen. Other calculators have slightly different mode options, so their screens will look different.)

The first two lines of this display control the way numbers are displayed on the home screen. On the first line highlight **Normal** for now. The **Sci** option refers to scientific notation, which you will not use in the IMP curriculum until Year 2. (*Note:* If the calculator displays a very large or very small number, it will use scientific notation no matter what option you highlight in the MODE screen.) The **Eng** option refers to engineering notation, which is a variation of scientific notation. You will not need this option in IMP.

On the second line highlight the **Float** option. When this option is in use, the calculator displays numbers using up to 10 digits, as well as a negative sign and a decimal point (if needed).

The third line refers to units for angle measurement. Highlight **Degree**. You will not need radian measurement of angles in IMP until Year 4.

Some of the other standard mode settings are **Func**, **Connected**, **Sequential**, **REAL**, and **FULL**. Highlight these if you have not already done so.

Finding the Square Root

`√(4*2+1)`

To calculate a square root, press [2nd] [√]. In order to find the square root of a long expression, use parentheses. The square root function includes an open parenthesis, so you should close the expression within the radical with the close parenthesis.

`3+4!+1-2`

Using the Factorial Command

To find the factorial symbol, press MATH, highlight the **PRB** (probability) menu heading, and find ! in the submenu.

`((1+4)*3)²`

Using Parentheses as Grouping Symbols

In printed mathematical expressions, you can use parentheses ( ), brackets [ ], and braces { } as grouping symbols. However, your TI calculator only uses parentheses as grouping symbols. Brackets and braces are used for other kinds of mathematical notation.

For example, to evaluate the expression $[(1+4) \cdot 3]^2$ on your calculator, you must replace the brackets with parentheses and enter $((1+4) \cdot 3)^2$.

Using Lists to Build a Spreadsheet

To work with a list, press $\boxed{\text{STAT}}$. **EDIT** is already selected, so just press $\boxed{\text{ENTER}}$. You should see at least three lists of numbers. For now, you are going to use list L1 and list L2. If your lists already contain values, you will want to clear them first. (If others have been using the calculator, you might check with them before you clear their lists.) The fastest way to clear a list is to put the cursor on the list name (in this example the cursor is on L2) and then press $\boxed{\text{CLEAR}}$ $\boxed{\text{ENTER}}$.

List L1 will contain the number of bagels Marcella has when she gets home. List L2 will contain the number of bagels Marcella started with. In the original example Marcella had no bagels when she got home, so the first number in list L1 will be 0. Move the cursor to the first entry in list L1 and press $\boxed{0}$ $\boxed{\text{ENTER}}$. Continue to the numbers 1, 2, 3, and so on until you get at least as far as 10.

Now create list L2. You could figure out these entries and enter them one at a time, but instead you will use your understanding of variables and algebraic expressions to make the calculator build this list for you.

You have discovered that in order to generate a number in list L2, you take the corresponding number in list L1, multiply it by 8, and add 28 to that result. To tell your calculator to do exactly that for the entire list, put your cursor on the label L2. Now enter 8*L1+28. (Find L1 by pressing $\boxed{\text{2nd}}$ [L1].) Press $\boxed{\text{ENTER}}$ when you are done. Use your arrow keys to examine the entries in your completed list 2. If Marcella ends up with 5 bagels, how many did she start with?

If you change the entries in list L1, you'll see that the entries in list L2 do not change. That's because your calculator does not remember that list L2 is calculated by 8*L1+28. If you want list L2 to automatically update when you change list L1, you can permanently attach the expression to the list using quotation marks, $\boxed{\text{ALPHA}}$ ["]. Now the list has a "lock" symbol and its entries update when you change entries in list L1.

The "lock" symbol.

Building a Sequence

You can use your calculator to create a sequence. Your sequence will appear in a list. These instructions use an example from *Add It Up* and work with the sequence that follows the rule $4t^2 + 3$, starting with $t = 5$. To demonstrate the power of the calculator, let's have the sequence end with $t = 50$. (Later on we will add all the terms of this sequence.)

First, press [STAT] [ENTER] to show your lists. You will build your sequence in list **L1**, so place your cursor on **L1**. Clear this list if needed by pressing [CLEAR] [ENTER]. Put your cursor back on **L1** if necessary, and press [2nd] [LIST]. In the **OPS** menu highlight **seq(**. Press [ENTER] and return to your list.

After the **seq(** function enter **4X²+3, X, 5, 50, 1)**.

The information in parentheses gives your calculator the following information in order: the algebraic expression that describes your sequence, the variable in the expression, the lower and upper limits of the variable, and finally the increment for the variable x. It's easier for the calculator to use x as the variable in this case. That's why you're entering x instead of t.

Press [ENTER]. Use the arrow keys to navigate around your list.

Finding the Sum of a Sequence

```
sum(L1)
          171718
```

Once you have your sequence in a list, finding the sum is not difficult. Press 2nd [QUIT] to return to your home screen. Then press 2nd [LIST], but this time choose the **MATH** menu. Find the **SUM** function and press ENTER 2nd [L1]. Then press ENTER. Your calculator will display the sum of all the numbers in your list!

Now use your calculator to sum another sequence of numbers.

Programming Your Calculator for Borders

For the supplemental activity *Programming Borders* you will program your calculator to help the employees at an outdoor supply store. Your program will use information about the dimensions of a rectangular garden and calculate the number of border tiles, amount of topsoil, and cost.

The steps here tell you how to create a simpler program that uses the rule $4s - 4$ to calculate only the number of tiles needed for a one-tile-wide border around a square garden. After writing this simple program, you can embellish for rectangular gardens and thicker borders. (You could load the program **BORDER1.8xp** on your calculator.)

```
EXEC EDIT NEW
1:DOUBLES
2:*RANGER2
```

Writing the Program

Press PRGM to get a display like the first one shown here. The screen shows a list of preexisting programs on your calculator, if any.

```
EXEC EDIT NEW
1:Create New
```

Highlight **NEW**, and press ENTER.

```
PROGRAM
Name=BORDER■
```

To name your program, type **BORDER** and press ENTER. In this screen, the Alpha-Lock is on by default, so you can just press the necessary letters.

You will actually write your program in the next screen. Certain keys perform differently when you are in the programming editor.

Press PRGM to see the program menu. Highlight **I/O**, choose **Input**, and press ENTER.

```
CTL I/O EXEC
1:Input
2:Prompt
3:Disp
4:DispGraph
5:DispTable
6:Output(
7↓getKey
```

Continued on next page

```
PROGRAM:BORDER
:Input S
:
```

The **Input** instruction will appear on your programming screen. You want the program to store the input value for the variable **S**. Press [ALPHA] [S] to complete the program line as shown here. Press [ENTER] to go on to the next line of the program.

```
PROGRAM:BORDER
:Input S
:4S-4→K
:
```

Now enter the instruction **4S−4→K**. The symbol → means "store"; you enter it by pressing the [STO▸] key. This computes the value of the expression **4S−4** for the value of **S** that was previously input, and it stores that result in a variable called **K**. In other words, the value of **K** is the number of tiles needed. Press [ENTER] to go on to the next line of the program.

```
PROGRAM:BORDER
:Input S
:4S-4→K
:Disp K
```

Press [PRGM], highlight **I/O**, choose **Disp** ("display"), and press [ENTER]. Tell the calculator what to display by entering [ALPHA] [K]. This is the end of your program.

Running the Program

```
EXEC EDIT NEW
1:BORDER
2:DOUBLES
3:*RANGER2
```

To run your program, press [2nd] [QUIT] to return to your home screen. Then press [PRGM]. Choose **EXEC** ("execute"), highlight the **BORDER** program, and press [ENTER]. **PrgmBORDER** will appear on the home screen.

```
PrgmBORDER
?■
```

Press [ENTER] to run the program. The calculator will show **?** and a flashing cursor, indicating that you need to enter a number as an input.

Input any number and press [ENTER]. The calculator will display the output according to the rule $4s - 4$, For example, if you input the number 10, the calculator will compute $4(10) - 4$ and display the result, 36, as shown here. It will also tell you that the program is finished, by displaying **Done**.

```
PrgmBORDER
?10
                36
             Done
```

Continued on next page

To run your program again, you can press [PRGM] and choose **BORDER** again. However, if you haven't yet performed any other tasks on your calculator, simply press [ENTER], and the program will repeat. Test your program a few times using different input values.

Editing the Program

To change your **BORDER** program, press [PRGM], highlight **EDIT**, and choose **BORDER**. Use the arrow keys to move around within your program and edit, insert, or delete whatever words or commands you wish. To make the program fully functional for an outdoor supply store, you'll ultimately need three input variables and three output (display) variables. This happens in the program **BORDER3.8xp**.

Making the Program User-Friendly

```
prgmBORDER
SIDE LENGTH?
?10
TILES:
            36
          Done
```

It is always nice to include messages so that a user knows what's happening as the program runs. For example, the simple program can be modified to include input and display messages, as shown here. (See the program **BORDER2.8xp**.)

To do this, go back into program editor mode and insert or edit **Disp** lines. You'll probably want to turn on the Alpha-Lock. Each text message must be in quotation marks so it isn't confused as a variable. When a display involves both text and a variable, use a comma in between. For spaces between words, use [ALPHA] [#] above the zero key.

```
PROGRAM:BORDER
:Disp "SIDE LENG
TH?"
:Input S
:4S-4→K
:Disp "TILES:",K
```

Note: A message longer than 16 characters will not display completely on the home screen. If you want to include a longer message, break it into pieces and use several **Disp** commands. You might also want to research the **Output(** command, which allows you to display text or values at a specific location on the screen, and **ClrHome(**, which clears the home screen before displaying information.